Environmental Chemistry

Second edition

Peter O'Neill
University of Plymouth
Department of Environmental Sciences
Plymouth
UK

CHAPMAN & HALL
London · Glasgow · New York · Tokyo · Melbourne · Madras

Published by Chapman & Hall, 2–6 Boundary Row, London SE1 8HN

Chapman & Hall, 2–6 Boundary Row, London SE1 8HN, UK

Blackie Academic & Professional, Wester Cleddens Road, Bishopbriggs, Glasgow G64 2NZ, UK

Chapman & Hall Inc., 29 West 35th Street, New York NY 10001, USA

Chapman & Hall Japan, Thomson Publishing Japan, Hirakawacho Nemoto Building, 6F, 1-7-11 Hirakawa-cho, Chiyoda-ku, Tokyo 102, Japan

Chapman & Hall Australia, Thomas Nelson Australia, 102 Dodds Street, South Melbourne, Victoria 3205, Australia

Chapman & Hall India, R. Seshadri, 32 Second Main Road, CIT East, Madras 600 035, India

First edition 1985

Reprinted 1990, 1991

Second edition 1993

© 1985, 1993 Peter O'Neill

Typeset by Thomson Press (India) Ltd, New Delhi

Printed in Great Britain by St Edmundsbury Press, Bury St Edmunds, Suffolk

ISBN 0 412 48490 0

A catalogue record for this book is available from the British Library

Library of Congress Cataloging-in-Publication data

O'Neill, Peter.
    Environmental chemistry / Peter O'Neill. – 2nd ed.
       p.   cm.
    Includes bibliographical references and index.
    ISBN 0–412–48490–0 (pbk. : alk. paper)
    1. Environmental chemistry.   2. Inorganic compounds–Environmental aspects.   I. Title.
TD193.054   1993
574.5'222–dc20   92-46390
                                             CIP

♾ Printed on permanent acid-free text paper, manufactured in accordance with the proposed ANSI/NISO Z 39.48-199X and ANSI Z 39.48-1984

Contents

Preface

Since the first edition of this book was written knowledge of the Earth's chemical environment has increased. Yet there is still great uncertainty about the future changes that may result from human activities. It has become increasingly clear that these activities have the capacity to cause large-scale disruption of the natural environment, but whether that capacity will become a reality is a matter of dispute. The increase in the world's population and the increased mobilization of chemical species are producing global as well as local changes to the environment. Some of these problems are discussed in greater depth in this edition than in the first one. For example, the sections on greenhouse gases, ozone depletion and ground-level atmospheric pollution have all been extended. New chapters on radon and problems posed by persistent organic chemicals have been added, and there has been a general updating of information on all of the other elements.

Though the majority of this new material is connected with pollution, it must still be emphasized that environmental chemistry is much more than the study of pollution. Only by understanding the natural mobility of the elements and their compounds can the changes that human activities bring about be appreciated.

The breakdown of rock to form soils, the uptake of the mobilized chemicals by plants, and the return of the dead plant material to the soil ready for further uptake has long been known as a 'biogeochemical' cycle (indicating the interaction of biology, geology and chemistry). The biogeochemical cycle is only one part of the general geochemical cycle in which material is moved from the land to the sea, possibly having entered the atmosphere, and then being reincorporated in the land mass. The elements move through their cycles in fits and starts, with many variations in chemical form occurring along the way. Environmental chemistry attempts to explain why a specific change occurs and why a particular pathway has been followed: of necessity, there is overlap with biology, geology and physics. One possible definition of environmental chemistry is the study of the role of chemical elements in the synthesis and decomposition of natural materials of all kinds, including the changes specifically brought about by human actions.

The selection of topics for inclusion in this book proved difficult. The guiding principle has been to provide a broad survey illustrating the operation of natural systems, with some diversions to show how human activities can modify these systems. A four-part format has been chosen to allow the grouping together of related environmental topics. Most readers will have at least a knowledge of chemistry, but the text should be understandable even to those without this. A glossary giving brief explanations of terms and concepts used in the text is included. **Bold type** is used to indicate first appearances of terms that are to be found in the glossary or that are explained in the text. The book will have achieved its objective if, after reading it, you have gained some insight into the operation of chemical processes near the Earth's surface, and discovered that environmental chemistry is an exciting area of study.

The students for whom this book is designed will be taking first-year degree courses in environmental science or the various types of modular degree courses involving chemistry, geology, biology, ecology and physical geography options. In addition some second- and third-year special courses will find material of interest, as will science sixth-formers.

Part A sets the scene and provides an introduction to many of the basic geological, geochemical and chemical ideas that are essential for an understanding of geochemical cycles. The importance of oxygen to the chemistry of reactions occurring near the Earth's surface is emphasized by dealing with this element first. This importance applies to both biologically mediated and inorganic reactions; systems depending upon these reaction types are discussed in Parts B and C. Both differences and similarities between animate and inanimate systems are reviewed; above all, the application of general chemical concepts to what might appear to be widely different reactions is illustrated. Because the mobility of one element depends upon the other chemical species in a particular environment, different aspects of certain topics of concern (such as 'acid rain') are discussed in several chapters. In Part D there is a brief examination of some of the effects of human activities on elements that usually cycle naturally in small quantities. This final part further highlights the perturbation to natural cycles caused by agricultural, industrial and social developments with the consequent problems of environmental management.

I would like to thank all those students and staff who have read the manuscript either in full or in part. Their comments have proved very helpful, as has the advice and guidance I have received from the staff of the Publishers. However, the responsibility for the approach chosen and the views expressed remains with me.

P. O'Neill

Part One

The oxygen-rich planet

Although it could be argued that each planet in the solar system has its own unique features, the dioxygen-rich atmosphere and the presence of living matter make the Earth especially interesting. It was the presence of water rather than oxygen that was crucial to the development of life on Earth, but the present composition of the atmosphere ensures that organisms that can make use of the oxygen in energy-releasing reactions are dominant at the Earth's surface.

In Part A we shall look at (a) the geological development of the Earth, (b) the development of life-forms, and (c) the part oxygen has played in these developments. The presence of free or combined oxygen dominates the surface chemistry of the Earth (Table P1.1); for this reason oxygen is discussed in this introductory part of the book. Knowledge gained about oxygen's chemistry can then be applied when discussing the chemistry of the other major elements in living systems (Part Two) and of the other major elements in the Earth's crust (Part Three).

Any study of environmental chemistry soon reveals the dependence of the reactions of any one element on the reactions of a number of other elements. Therefore, although some reactions of oxygen are considered in Chapter 2, more detail is often given in later chapters when other elements are being investigated. For instance, the problems of **acid rain** involve the reactions of oxygen with compounds of sulphur and nitrogen to form products that dissolve in water to form acidic solutions. These solutions can interact with a wide variety of soils, rocks, buildings and so on, as well as affecting plants, animals and fish. As a consequence various aspects of acid-rain production and its effects are discussed in Chapters 2–6 and 10.

Chemical reactions involve the production of new distributions of the **electrons** among the atoms taking part in the reaction, or a change in the distribution of the atoms (Appendix). Lead in water pipes will

Table P1.1   The ten major elements in the Universe, the Earth, the Earth's crust, the ocean, the atmosphere as dry air, and the biosphere (as weight %)

Universe	Earth	Crust	Ocean	Atmosphere	Biosphere
H	Fe	O	O	N	O
77	35	46.6	85.8	75.5	53
He	O	Si	H	O	C
21	29	29.5	11	23.2	39
O	Si	Al	Cl	Ar	H
0.8	14	8.2	1.94	1.3	6.6
C	Mg	Fe	Na	C	N
0.3	14	5.0	1.05	9.3×10^{-3}	0.5
Ne	S	Ca	Mg	Ne	Ca
0.2	2.9	3.6	0.13	1.3×10^{-3}	0.4
Fe	Ni	Na	S	Kr	K
0.1	2.4	2.8	0.09	0.45×10^{-3}	0.2
Si	Ca	K	Ca	He	Si
0.07	2.1	2.6	0.041	72×10^{-6}	0.1
N	Al	Mg	K	Xe	P
0.06	1.8	2.1	0.039	40×10^{-6}	0.1
Mg	Na	Ti	Br	H	Mg
0.06	0.3	0.57	0.007	23×10^{-6}	0.1
S	P	H	C	S	S
0.04	0.2	0.22	0.003	70×10^{-9}	0.07

slowly dissolve in the water as the atoms of metallic lead lose two electrons each (Eqn P1.1) to some chemical species in the water.

$$Pb_{(solid)} \longrightarrow Pb^{2+}_{(solution)} + 2e^- \qquad (P1.1)$$

The consequence of this simple change could be the poisoning of people drinking the water. The charged form of lead in solution is in a lower energy state than the lead in the metal. All reactions occur in such a way that the **energy** of the total system is lowered (Chapter 4). Our understanding of the factors controlling these changes in energy is expressed by the laws of thermodynamics (Table 4.1). These laws and the consequent mathematical equations allow us to predict how likely it is that a given chemical reaction will occur.

Unfortunately thermodynamics tells us nothing about how *rapidly* the reaction will occur. The rate at which a reaction occurs (the **kinetics** of the change) is of the utmost importance in environmental chemistry. Many of the controversies about the likely impact of the use of a particular process or chemical involve disputes about chemical kinetics. If an insecticide is sprayed on to an edible crop, there will be no danger to someone eating the crop a month after spraying if the insecticide decomposes to harmless products within 24 hours. However, if the decomposition is much slower a health hazard could exist. **Enzymes** speed up chemical reactions and all organisms depend upon enzymic reactions for their survival. Many harmful substances, such as lead, are toxic because they react with the enzymes inside organisms and usually slow down or stop essential reactions. The understanding of natural systems depends upon an understanding of both the thermodynamics and the kinetics of the system.

1 History of the Earth

The origin of the solar system, including the Earth, is a subject of some controversy. This is because the various hypotheses are based on our very incomplete knowledge of the solar system as it exists now. The further back in time we go the greater the assumptions and extrapolations that have to be made about the conditions that existed then. The evidence we have indicates that by about 4.5 billion years (4.5×10^9a) ago there was a planet with a solid crust approximately 150 million kilometres from the Sun. Since that time major changes have occurred at a much slower rate.

The major structural features of the Earth (Figure 1.1) have probably shown little change in the past 4 billion years. Life on Earth is confined to the regions close to the atmosphere/land/ocean boundaries, and these portions of the globe are in a state of continuing flux.

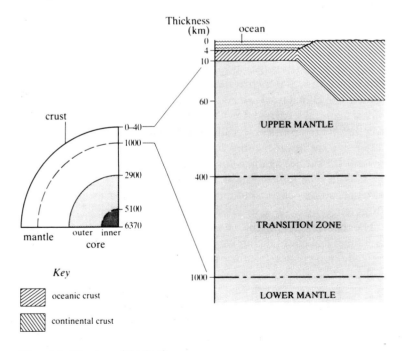

Figure 1.1 Structure of the Earth.

It is knowledge of the operation of this thin outer shell, which contributes less than 1% of the mass of the Earth, that is of such great importance to us if we are to ensure the survival of the human species and to attain a satisfactory quality of life.

The present surface of the Earth is irregular, about 70% being covered with water of average depth 4 km (range 0–11 km). The remainder comprises land masses with an average height of 0.84 km (range 0–8.8 km) above sea level. Two-thirds of this land is situated in the Northern Hemisphere. The major land masses are called **continents** and they are best defined not by shorelines but by the edges of the deep-sea platforms, which may be far out to sea; they thus incorporate the continental shelf (Figure 1.2). This definition of continents means that offshore islands are included as parts of their nearest land masses. More importantly, it agrees with the geological and chemical evidence that there is a compositional difference between continental crustal rocks and the crustal rocks underlying the deep oceans (Figure 1.2).

It is now thought that the Earth's surface consists of a number of rigid plates (100–150 km thick) moving discontinuously over the mantle (Figure 1.3). These plates consist of the continental and oceanic crusts plus the uppermost layer of the mantle. Where two adjoining plates are moving apart, at ocean ridges, the region is marked by volcanism, by the addition of new crustal material, and by the formation of oceanic basins, e.g. the Atlantic and Indian Oceans. When two plates converge, one plate passes partially under the other to form either an arc–trench system, in oceanic regions, or a mountain belt (e.g. the Himalayas), where two continental regions collide. When crustal material is carried down towards the mantle as one plate rides over the other, the process is called **subduction**. This material becomes hotter, and the lower-melting, lighter components may melt and rise towards the surface again, before the denser components mix in with the mantle. The reinjection of these large masses of lighter molten material starts off a mountain-building sequence (Figure 1.3). Mountain ranges are associated with the convergence of plates.

The relative direction of the movement of adjacent plates changes with time. The plates themselves also change as they break up or fuse with adjoining plates, but the overall effect is to keep the irregularity of the Earth's surface contours. A good example of change is along the Pacific coast of North America, where two converging plates, the collision of which formed the Rocky Mountain ranges, are now sliding past each other. The lateral movement has produced an extensive transform **fault** system, the best-known member of which is the San Andreas Fault. The effect of this movement is to separate the

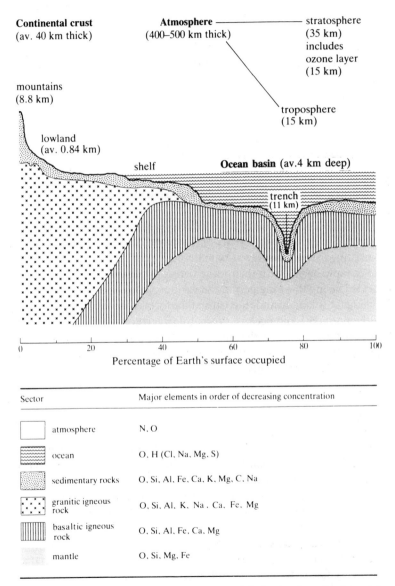

Figure 1.2 A cross-section of the surface of the Earth, indicating the major elements in various sectors.

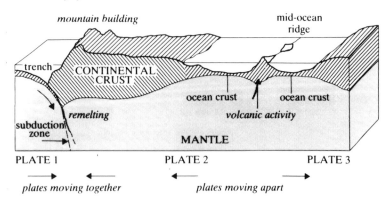

Figure 1.3 The formation of mid-ocean ridges as plates move apart, and of subduction zones and mountains as plates move together.

Californian coastal region from the mainland and move it northwards towards Canada.

The oldest dated rocks are 3.8 billion years old. It seems that for at least the past 4 billion years there have been continental areas consisting of less dense (approximately 2.7 g cm^{-3}) rocks rising above the oceanic basins with their more dense (approximately 3.2 g cm^{-3}) rocks. During this long period the various plates have moved, joined together, broken up and thus caused the formation and destruction of both continents and oceans. Present-day earthquakes and volcanoes are constant reminders that these large-scale processes are still continuing. However, most of this book is concerned with processes that occur much more rapidly.

1.1 Development of life

If it is true that the Earth's development included a period when the surface was molten, then life-forms of the types we know nowadays could not have existed at those temperatures. Our knowledge of life in the times before there were written records, or even before there were humans to write, depends upon a study of the preserved remains of organisms. These remains, called **fossils**, are found in sedimentary rocks; generally only the hard parts, e.g. skeletons and shells, are preserved. Only under exceptional conditions, usually associated with very fine sediments, do soft organs leave an imprint.

The oldest recognizable fossils are over 3 billion years old and are similar to some present-day bacteria and algae, i.e. single-celled

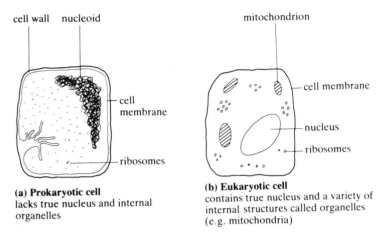

cell wall   nucleoid

cell membrane

ribosomes

mitochondrion

cell membrane

nucleus

ribosomes

(a) Prokaryotic cell
lacks true nucleus and internal
organelles

(b) Eukaryotic cell
contains true nucleus and a variety of
internal structures called organelles
(e.g. mitochondria)

Figure 1.4 The two major types of living cells: (a) prokaryotic cells; (b) eukaryotic cells.

organisms. The **cells** do not have well-developed nuclei and the organisms are therefore called **prokaryotes** (Figure 1.4). Though bacteria and algae are relatively simple life-forms, they have a high degree of order and are capable of carrying out complex biochemical reactions. The processes by which these prokaryotes could have developed are still clouded in mystery.

The most popular view of the development of life assumes that there was a gradual progression from simple **inorganic molecules** to the present-day wide diversity of simple and complex life-forms (Figure 1.5). All living things on Earth are composed of compounds containing strings of carbon atoms. Because many of them are essential to life, carbon-containing compounds and molecules are often called **organic** compounds and organic molecules (Part Two). A large number of organic compounds can be made in laboratories by reactions between water (H_2O), ammonia (NH_3), carbon dioxide (CO_2), carbon monoxide (CO), methane (CH_4), hydrogen sulphide (H_2S), and dihydrogen (H_2). These simple molecules are thought to have been present in the atmosphere and oceans of the primitive Earth. It seems reasonable to suppose that such reactions could have occurred and that a wide range of organic molecules could have been produced. It is the steps from non-living systems to self-supporting, self-replicating living organisms and the subsequent development of diverse life-forms that are most speculative.

The success of an **organism** (any living creature, or life-form) is dependent upon the efficiency with which it utilizes the available

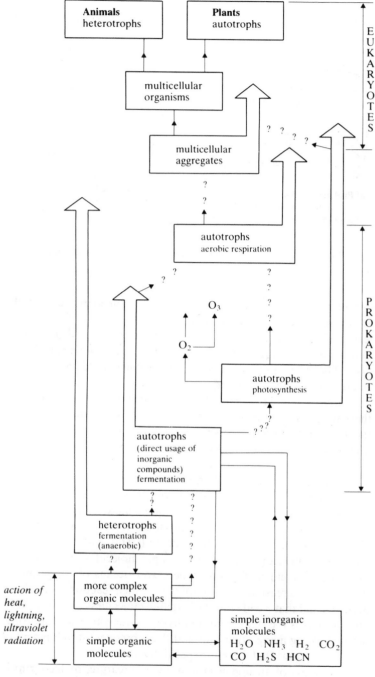

Figure 1.5 A possible pattern for the development of life on Earth.

sources of food and energy sources. The earliest organisms were dependent on the supply of *externally* synthesized organic molecules. Like humans and other animals, they were **heterotrophs**. **Autotrophs** are capable of synthesizing the required organic molecules from simple inorganic molecules. As these simple molcules were much more abundant in the ocean and atmosphere than were the organic molecules, the autotrophs may be expected to have expanded at the expense of the first heterotrophs.

Both types of prokaryotic organisms would obtain energy by **fermentation** reactions (Eqn 1.1).

$$C_6H_{12}O_6 \longrightarrow 2C_3H_4O_3 + 4H \qquad (1.1)$$

glucose   pyruvic acid   (combined with other groups)

Fermentation is not a very good source of energy. The ability to utilize the visible part of the spectrum, and so use **solar radiation** as an energy source for the conversion of carbon dioxide into organic molecules, gave the **photosynthetic** bacteria a big advantage. Initially molecules such as hydrogen sulphide, H_2S, or simple organic compounds were used as hydrogen donors (Eqn 1.2).

$$nCO_2 + 2nH_2A \xrightarrow{\text{light}} (CH_2O)n + nH_2O + 2nA \qquad (1.2)$$

carbon dioxide   hydrogen donor   carbohydrate

Then blue-green algae developed which could make use of the almost limitless water in the oceans as the hydrogen donor and which produced dioxygen as a byproduct (Eqn 1.3).

$$nCO_2 + nH_2O \longrightarrow (CH_2O)n + nO_2 \qquad (1.3)$$

carbon dioxide   hydrogen donor   carbohydrate   dioxygen

The release of dioxygen brought about major changes to the Earth's surface (Chapter 2), including the killing off of many organisms that were not able to protect themselves against this reactive gas.

As the level of dioxygen in the atmosphere built up, an ozone (O_3) layer gradually formed in the stratosphere (15–60 km above the Earth's surface). Ozone is very good at absorbing harmful ultraviolet radiation. As the level of ultraviolet radiation reaching the Earth's surface decreased, organisms could begin to colonize regions close to the atmosphere/water/land interface: there was no longer the need for a protective layer of water to filter out the ultraviolet. The presence of dioxygen allowed suitably modified cells to use **respiration** reactions (Eqn 1.4) to obtain 18 times more energy than the fermentation reaction releases.

$$(CH_2O)n + nO_2 \longrightarrow nCO_2 + nH_2O \qquad (1.4)$$

carbohydrate   carbon dioxide   water

The organization of the material inside the cells now underwent a significant change, with the development of (a) a membrane-bounded nucleus containing the nucleic acids carrying the cell's genetic information, and (b) a number of other distinct structural features (Figure 1.4). These new cells are called **eukaryotic** cells as they contain a definite nucleus. The autotrophic single-celled eukaryotes evolved to multi-celled green plants which used photosynthesis to produce organic molecules and dioxygen. The expansion in population of efficient photosynthesizing and respiring organisms provided the basis for the resurgence of heterotrophs. The eukaryotic heterotrophs evolved into the fish, insects and animals that are so abundant today.

Four billion years ago the Earth's atmosphere probably consisted mainly of dinitrogen (N_2), and possibly carbon dioxide (CO_2), with minor amounts of dihydrogen (H_2), carbon monoxide (CO), methane (CH_4) and possibly dioxygen (O_2). The development of life, as outlined above, has caused dioxygen to become a major component (21%) of the atmosphere, with carbon dioxide being a minor component (354 ppm; 'ppm' means 'parts per million'); and dihydrogen (0.5 ppm), methane (1.5 ppm) and carbon monoxide (0.1 ppm) becoming very minor components. Environmental change is a continuing process, and, as dioxygen-breathing heterotrophs, maybe we should not be too upset by this.

1.2 Distribution of the elements

If we wish to study the chemical changes that take place in the environment, it is important to have some sort of reference system. This reference system should allow us to recognize whether a chemical change is normal or abnormal for the particular environment being studied and what will be the size of the effect brought about by the change. In order to recognize the abnormal we must know what is normal, and to recognize change we must know what the original state was. These statements may appear to be self-evident, but to be aware of the environment in the necessary detail is extremely difficult.

A knowledge of the chemical composition of a system is necessary before we can meaningfully study the chemical changes going on in that system. When we look carefully at what we know about the chemical composition of the various components of our environment (Table P1.1), we realize how many estimates must be included in these figures. It is very easy to see tables of numerical values and assume that they are reliable representations of some information. If the information has been obtained by chemical analysis, the analysis is said to be 'accurate' if it gives the true value.

The **accuracy** of a method is affected by **errors**, both random and systematic. The smaller these errors, the more accurate and precise the results. **Precision** can be defined as the ability to obtain the same results – right or wrong – repeatedly. Precision is actually a measure of the random errors in a system. Systematic errors lead to an inaccurate result biased towards one value (Figure 1.6). 'High' or 'good' precision means that there is little variation in the values. The precision of a method can only be estimated by repeating the analysis of the sample many times. Statistical treatment of the results then allows a numerical value for the precision to be obtained, e.g. the percentage relative variation at the 95% confidence limit (this value is almost twice the **standard deviation**) is often used (Figure 1.7). If the determination of the iron content of a rock gives a value of 42.0 g kg^{-1} with a precision of $\pm 10\%$, then this means that we would expect 95 out of 100 analyses of the rock to give values in the range of 37.8–46.2 g Fe kg^{-1}, and five results to be higher or lower. If the distribution of results is

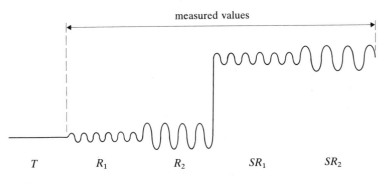

Key

T |true value

Measured values:

R_1 small error showing relatively high precision (i.e. good repeatability with little variation)

R_2 larger errors giving relatively low precision (i.e. poor repeatability with larger variation between single values)

SR_1 systematic error plus small random errors giving biased results; the bias could be due to contamination, faulty instrumentation, etc. – high precision

SR_2 systematic error plus greater random error again giving biased results – lower precision

mean of either R_1 or R_2 accords with T . the true value $\therefore R_1$ and R_2 provide good accuracy
neither SR_1 nor SR_2 give true values $\therefore SR_1$ and SR_2 provide poor accuracy

Figure 1.6   The effects of random and systematic errors on measured values.

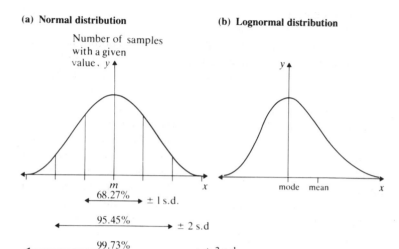

(a) Normal distribution   **(b) Lognormal distribution**

Proportion of results that are within 1 2 or 3 standard deviations (s.d.) of mean value *m*

Positively skewed results – if the logarithm of *x* is used the distribution becomes normal as in (a) and the mode (the most common value) becomes the mean (average) value

Figure 1.7   Normal and lognormal distributions of results.

normal (Figure 1.7) then the value of $42.0\,g\,Fe\,kg^{-1}$ is the **mean**, or arithmetic average, of all the results. In practice it is found that the elemental distributions in rocks are often lognormal, and it is the *logarithmic* values of the concentrations that are distributed normally (Figure 1.7).

Though many instrumental measurement techniques will give precisions of the order of 1%, pre-analysis treatment of rocks involving grinding, sieving, dissolution and so forth will decrease the precision, and values of 5–10% are often the best that can be attained. When the variability introduced by sample collection and storage is included, the precision can drop to 50% or less. The importance of determining precision is recognized much more nowadays, but results are still often reported with no indication of their reproducibility. However, we must always keep in mind the fact that systematic errors will not be picked up by simple precision tests and that highly precise results may be very inaccurate. Accurate results must be precise, but precise results may not be accurate!

Turning our attention to Table P1.1, it is obvious that the values given cannot be very precise. The concentrations of the ten major elements in the Universe and in the whole Earth must be based on few analyses and many assumptions; neither region is easily accessible for direct analysis. The oceans and the atmosphere are both fluids, but are stratified and not as homogeneous as might at first be thought. The concentrations of the major elements are reasonably well known. The minor elements are less well quantified because of their very low concentrations, with instrument **sensitivity** (ability to determine small amounts) and contamination becoming critical. The **biosphere** (all the organic matter in the atmosphere, oceans and crust) and the crust are both very heterogeneous so that sampling becomes the critical factor.

The crustal figures highlight the problems. First we must define the boundaries of the crust. The crust is the solid material above the **Mohorovičić discontinuity**, which is found at 5–6 km below the ocean bed and 40–60 km below the continental surface. Is the crust homogeneous? Quite clearly it is not. The crust is composed of many different rock types such as granites, basalts, sandstones and limestones, each of which has its own distinctive composition. In addition each rock type will vary in composition from place to place. A recent examination of one rock outcrop covered about 80 km^2 and involved analysing 93 samples: this gave the mean concentration of iron as being 4.75% with a standard deviation of 1.15% Fe, indicating that about 88 of the samples lie in the range 2.45–7.05% Fe. This wide range is not unusual. Rocks at the surface are accessible, but how do we collect samples from 40 km or 50 km below the surface? It is not surprising that different estimates of crustal abundances give different values. The estimates are based upon the assumptions that the analyses are representative of the various rock types and that the correct distribution of rock types, both in area and depth, is chosen. Values given to more than two **significant figures** imply a spurious degree of precision.

1.3 Geochemical cycles

Tables such as Table P1.1, showing the distribution of the elements, ignore the movement of chemical substances from one of the chosen divisions to another. Rivers carry dissolved and suspended material from the land to the oceans. Movement of plates causes the uplift of ocean sediments and the formation of new land masses. Plants add dioxygen to the atmosphere by photosynthesis and remove it by respiration. All of these movements might be described as natural because they occurred before humans appeared on the Earth. The presence and activities of the human population have caused changes in the rates of movement of many chemical substances and it is the possible effects of these changed rates that have been causing increasing concern. Not only are there increases in transfer rates of compounds already present (such as carbon dioxide, which is being released by the burning of carbon-containing fuels), but completely new chemicals (such as the insecticide DDT) are being distributed on land and sea. The flow of a chemical element between land mass, ocean and atmosphere can be described by means of a geochemical cycle.

Each geochemical cycle is a model that describes the movement of a chemical element or species, usually near the surface of the Earth. The vast majority of material now near the Earth's surface – the crust, the ocean and the atmosphere – stays in this region. There appears to be little loss either to the mantle or to outer space. Because the movement is both continuous, even if very slow in some cases, and conservative (there are few losses or gains), there is a repeated cycling of the components. The cycles are described using reservoirs, physically well-defined units, and transport paths along which material is transferred from one reservoir to another (Figure 1.8). The number of reservoirs used depends upon the degree of detail with which the system is to be studied. The degree of detail that is actually possible is determined by the amount and quality of the data available on the reservoir contents and transport pathways. One of the major consequences of the attempts to construct balanced geochemical cycles has been the highlighting of those parts of the cycle about which we have very little knowledge.

In quantitative geochemical cycles estimates are made of both the amount of an element in each reservoir and the flux between the reservoirs. The flux is defined as the amount of material passing along a particular transport pathway in a fixed period of time, usually 1 year (Figure 1.8). If the concentration of an element remains constant in a reservoir this means that there is a balance between the amount of element entering the reservoir and the amount leaving. A steady state has been reached. Most natural cycles, i.e. those not influenced by human technology, appear on a global basis to be in a steady state. This conclusion has been reached because there is little geological evidence to show that there has been any great change in the global chemical environment over the past few hundred million years.

If we know that a system is in a steady state, we can determine the residence times for individual elements in particular reservoirs (Eqn 1.5).

$$\text{residence time} = \frac{\text{amount of element in reservoir}}{\text{rate of addition (or removal) of the element}} \quad (1.5)$$

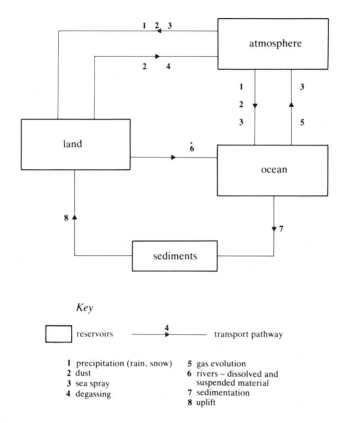

Figure 1.8   A general model of the geochemical cycle.

For example, if the amount of sodium dissolved in the ocean is 15×10^{18} kg, and the amount added each year is 100×10^9 kg , then the residence time will be 150 million years. (Sodium actually has a residence time of 210 million years.) Most of the elements have residence times in the ocean of a few million years or less. Residence times in the crust are usually much longer and residence times in the atmosphere are shorter, reflecting the mobility of the systems.

Possibly the largest impact humans have had on general geochemical cycling processes has been the increase in the rate of transfer of solids from land to sea. This has mainly occurred by disturbing the plants that hold soil in place. In forest areas the cutting down of trees has allowed the rapid washing away of soil and loose rock, whereas in other areas bad agricultural practices have caused soil to be washed away and blown away. It is thought that land-erosion rates are now at

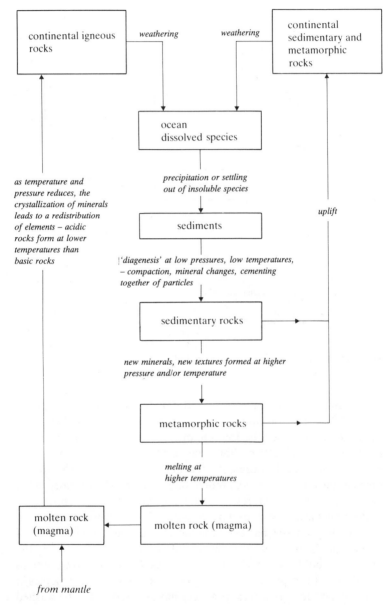

Figure 1.9 Geological processes involved in geochemical cycles.

least double what they were 5000 years ago. This degree of interference with the system we are measuring makes monitoring man's effects very difficult as we can not be certain about what our base-line concentrations and fluxes should be.

The general geological processes involved in the geochemical cycling of the elements are indicated in Figure 1.9. The changes in rock types are brought about by the changes in stability of the individual chemical components under different conditions of temperature and pressure. In addition, the presence or absence of a particular chemical species influences the chemical reactions that occur. The presence of both water and dioxygen at the Earth's surface has a dominating effect on the reactions occurring in this zone. Many of these reactions are also influenced by biological activity, and the interaction of chemical and biological systems will be a major theme in the following chapters.

Though a knowledge of global geochemical cycles is important we are also interested in what happens on a much smaller scale. The global average can hide very wide local fluctuations. The leaching of cadmium from a mine dump, which may cause fish downstream to be killed, is an event of local importance; but this extra flow of cadmium is negligible in comparison with the total global cycle. We therefore need models that will allow us to illustrate and predict what will happen in local landscape units.

Landscape geochemistry is an holistic approach that looks at local events. It has been described as 'the study of the role chemical elements play in the syntheses and decomposition of natural material of all kinds' (J. A. C. Fortescue). If we add, 'the study of the changes specifically brought about by human actions', we have a good definition of environmental chemistry. In the ecosystem approach, the living biosphere is the central theme and the behaviour of chemical elements in the environment is of secondary importance. In environmental chemistry, the behaviour of the chemical entities is of central importance. Living or dead matter, along with pedological (affecting soils) and geological (affecting rocks) processes, are considered as contributing factors to the circulation of chemical elements.

xygen

Oxygen Abundance by weight (the relative abundance is given in parentheses): Earth, 28.5% (2); crust, 46.6% (1); ocean, total 85.8% (1), dissolved O_2 6 ppm at 15 °C (13); atmosphere, 23.2% (2). ppm = $mg\,kg^{-1}$.

Oxygen has an uniquely important place in the composition of both the animate and inanimate world. It is the only element that is present in high concentrations in the crust, atmosphere, hydrosphere and biosphere. For this reason oxygen deserves special consideration.

The major rock-forming minerals in the Earth's crust are **silicates**, whose main structural unit is the $[SiO_4]$ tetrahedron (Chapter 8). About 95% of all crustal rocks are composed of silicate minerals, and most of the remaining 5% is composed of minerals containing oxygen, e.g. as carbonate (CO_3^{2-}) in limestones, as sulphate (SO_4^{2-}) in evaporites, and as phosphate (PO_4^{3-}) in phosphate rock. The oxygen when combined in these minerals has an ionic radius of 140 pm (pm = picometre = 10^{-12} m), whereas the silicon has a radius of 26 pm, and all the other major elements, apart from calcium (100 pm), sodium (102 pm) and potassium (138 pm), have radii below 80 pm. As a consequence of the relatively large size of oxygen ions, about 94% by volume of the Earth's crust is oxygen.

The various oxygen-containing species such as silicates (SiO_4^{4-}) and carbonates (CO_3^{2-}) are called 'oxyanions' because they have a negative charge (a property of **anions**) and contain oxygen. Even when the mineral breaks down during weathering, the oxyanion usually remains as an unbroken unit that is transported to the ocean, incorporated into sediments, and returned to the land unaltered. The majority of the oxygen in the Earth's crust can be considered to be chemically inert, as it remains bound to the same central atom in an oxyanion and undergoes the minimum of chemical change as it passes through the geochemical cycle (Figure 2.1).

In contrast, the oxygen in the atmosphere is chemically reactive. The oxygen is mainly present as the free element, in the form of the dioxygen molecule, O_2. The reactivity of dioxygen is great enough to have a controlling influence on the geochemical cycles of many other elements, such as carbon, hydrogen, nitrogen, sulphur and iron. As

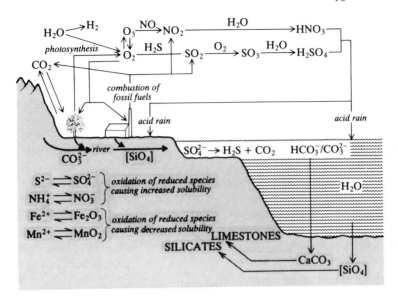

Figure 2.1 Chemical species found in the oxygen cycle.

has already been mentioned, the majority of the dioxygen in the atmosphere has been produced by photosynthesis. In addition, a small amount is produced by the action of ultraviolet (UV) light on water molecules in the upper regions of the atmosphere. This reaction (Eqn 2.1) is called **photodissociation**, and the hydrogen produced escapes into space.

$$2H_2O \xrightarrow{\text{UV light}} 2H_2 + O_2 \tag{2.1}$$

Ultraviolet radiation is also involved in the conversion of dioxygen into the triatomic molecule ozone, O_3. This reaction (Eqn 2.2) is called **photolysis** and is of particular importance because of the ability of ozone to absorb ultraviolet radiation.

$$O_2 \xrightarrow{\text{UV light}} O + O \,(\text{atomic oxygen}) \tag{2.2}$$

$$O + O_2 \longrightarrow O_3 \tag{2.3}$$

This absorption prevents the majority of the energetic ultraviolet radiation that is lethal to most life-forms from reaching the Earth's surface.

Respiration (Eqn. 2.4) is used by the majority of organisms as a means of producing energy; this reaction, together with the oxidation of dead organic matter (a chemically similar reaction) reverses the photosynthetic process.

$$(CH_2O)n + nCO_2 \longrightarrow nCO_2 + nH_2O \qquad (2.4)$$

carbohydrate   dioxygen   carbon dioxide   water

The burning of fossil fuels also produces carbon dioxide and water at the expense of dioxygen, but this is just a delayed **oxidation** of dead organic matter. Hydrogen sulphide, H_2S, from volcanoes and organic decay, is converted to sulphur dioxide, SO_2, in the atmosphere. This sulphur dioxide, plus that produced by the combustion of fossil fuels and the extraction of metals from sulphide ores, is eventually converted to sulphate, SO_4^{2-} (Figure 2.1), and returns to the Earth's surface, often as acid rain. Similarly, the various oxides of nitrogen, whether produced by micro-organisms or by human activities, are eventually converted to nitrate, NO_3^-, and may also produce acid rain. The phrase 'anthropogenic sources' is often used to indicate that something has been produced by human activities. On land, and to a lesser extent in the oceans, many metals are converted to insoluble oxides (Eqn 2.5) by atmospheric dioxygen, but some other **reduced** species (usually compounds of non-metals) are oxidized to more soluble forms (Eqn 2.6).

$$4Fe^{2+} + 3O_2 \longrightarrow 2Fe_2O_3 \qquad (2.5)$$

reduced   iron (II)   iron (III) oxide   **oxidized**
forms   (soluble)   (insoluble)   **forms**

$$S^{2-} + 2O_2 \longrightarrow SO_4^{2-} \qquad (2.6)$$

sulphide   sulphate
(insoluble)   (soluble)

At present there seems to be a state of equilibrium between the rate of formation and the rate of utilization of dioxygen in the atmosphere. This implies that there are fairly rapid feedback mechanisms in the system which counterbalance any changes in rates of formation or removal. The feedback mechanism is probably linked to the carbon cycle and the amount of organic matter incorporated in the ocean sediments. If the concentration of dioxygen drops, more carbon is incorporated in the sediments, i.e. photosynthesis is increased compared to removal processes; if the concentration of dioxygen rises, less carbon is incorporated in the sediments, i.e. removal processes are increased relative to photosynthesis.

2.1 Molecular oxygen in the atmosphere

Looking at Figure 2.1, we see that carbon dioxide (CO_2), sulphur dioxide (SO_2) and nitric oxide (NO) form components that, like elemental oxygen, are gases at room temperatures and atmospheric pressures. These gases are all composed of small molecules. A **molecule** contains a fixed number of specific types of atoms. It has characteristic properties: if there is a change in the number, or type, of any of the component atoms, there will be a change in the properties and it will have become a different molecule. Molecules have very strong internal forces holding together the atoms of which they are made (i.e. strong inter-atomic forces), but the forces between one molecule and the next are much weaker (i.e. weak inter-molecular forces). In general we find that the greater the mass of the molecule, the greater the energy required to break up the ordered structure of a solid (to form first a liquid, then a gas). The gases mentioned above are all composed of molecules containing a small number of light atoms and thus they have small relative molecular mass, e.g. $CO_2 = 44$, $NO = 30$, $O_2 = 32$, $SO_2 = 64$. One outstanding exception is water, which has a lower relative molecular mass (18) than these gaseous compounds, yet exists as a liquid under temperature conditions on the Earth's surface. The reasons for this fortunate divergence from the norm will be considered in Chapter 3.

Dioxygen

The presence of dioxygen in the atmosphere means that the reactions occurring on the Earth's surface are mainly carried out under oxidizing conditions. One of the pieces of evidence that suggests that the early atmosphere did not contain dioxygen is the presence of pyrites, FeS_2, in sedimentary rocks formed by deposition in contact with the atmosphere. Nowadays oxidation reactions (Eqns 2.7 and 2.8) tend to occur. These two equations could be combined into one (Eqn. 2.9) to give the overall reaction.

$$2FeS_2 + 7O_2 + 2H_2O \longrightarrow 2Fe^{2+} + 4SO_4^{2-} + 4H^+ \qquad (2.7)$$

$$4Fe^{2+} + O_2 + 4H^+ \longrightarrow 4Fe^{3+} + 2H_2O \qquad (2.8)$$

$$4FeS_2 + 15O_2 + 2H_2O \longrightarrow 4Fe^{3+} + 8SO_4^{2-} + 4H^+ \qquad (2.9)$$

An early definition of oxidation was that it involved the addition of oxygen to the substance being oxidized. In Equation 2.7 we see that the sulphur in the pyrites becomes oxidized to sulphate. However, it became apparent that the addition of oxygen was only one type of a much larger group of reactions that involved the transfer of electrons from one substance to another. In all these reactions the species that donates the electrons is described as being **oxidized**. In Equation 2.8, the iron(II), or ferrous, iron (Fe^{2+}) loses an electron to become iron(III), or ferric, iron (Fe^{3+}). The iron is said to have been oxidized.

If electrons are being lost by some atoms during oxidation, then other atoms must be taking up these electrons. Atoms that gain electrons are said to be **reduced**. Oxidation and reduction *must* occur in association, so that one set of atoms donates electrons and another set of atoms accepts electrons. Therefore there can only be combined reduction–oxidation reactions, which are usually called **redox reactions**. Sometimes it is convenient to consider only the oxidation half-reaction (Eqn 2.10) or the reduction half-reaction (Eqn 2.11), but neither of these half-reactions can occur by itself.

$$\text{oxidation half-reaction } Fe^{2+} \xrightarrow{\text{electron released}} Fe^{3+} + e^- \qquad (2.10)$$

$$\text{reduction half-reaction } O_2 + 4e^- \xrightarrow{\text{electrons accepted}} 2O^{2-} \qquad (2.11)$$

The oxidation half-reaction must be paired up with an electron acceptor and the reduction half-reaction must be paired up with an electron donor.

A redox reaction will occur when there is both a substance that will accept electrons (an **oxidizing agent**) and a substance that will donate electrons (a **reducing agent**). The greater its ability to attract electrons, the stronger the oxidizing agent is; and the greater the likelihood that a redox reaction will occur. Similarly, the greater the ability to donate electrons, the stronger the reducing agent. The tendency to accept electrons can be measured, and this measured electrode potential is often expressed as the standard **reduction potential**, E^0. The electrode potentials allow us to compare oxidizing and reducing agents, and gives a measure of their relative strengths. The actual reduction potential varies with the temperature and concentration of reactants, so this must be taken into account when investigating environmental reactions.

A drawback to lists of reduction potentials is that they indicate whether a reaction is theoretically possible, but they do not indicate the rate of the reaction. This kinetic problem is the reason for the controversy over the interpretation to be put on the presence of pyrites in ancient sediments. The electrode potentials and experimental evidence indicate that pyrites will be oxidized by moist dioxygen, but we also know that samples of pyrites may be preserved unaltered by dioxygen for long periods if kept dry. Was there no dioxygen present when the sediments were deposited, or was dioxygen present but without the other conditions required for rapid oxidation?

There has been an international agreement to express electrode potentials in terms of the standard reduction potentials, but for many environmental problems the term **oxidation potential** is used. The oxidation potential is just the reverse of the reduction potential. The signs of the potential are also reversed; e.g. the standard reduction potential for lead ($Pb^{2+} + 2e^- \longrightarrow Pb$) is -0.126 V and the oxidation potential is $+0.126$ V ($Pb \longrightarrow Pb^{2+} + 2e^-$).

We shall come across many redox reactions in the study of environmental systems. The changes in solubility that often accompany these changes in oxidation state can play a major part in controlling the mobility of many elements.

Photosynthesis

Photosynthesis is arguably the most important step in the oxygen cycle. Not only is dioxygen produced – dioxygen which is used by humans and other animals to liberate the energy necessary for life by oxidizing organic chemicals – but the organic chemicals themselves are derived from the carbon consumed in the photosynthesis reaction (Eqn 2.12).

$$nCO_2 + nH_2O \xrightarrow{\text{light}} (CH_2O)n + nO_2 \qquad (2.12)$$
$$\text{carbon dioxide} \quad \text{water} \qquad \qquad \text{carbohydrate} \quad \text{dioxygen}$$

Water is the most common hydrogen donor, with the hydrogen being transferred to the carbon atoms obtained from the carbon dioxide, and the oxygen from the water being liberated as dioxygen. The photosynthetic reaction only occurs in the presence of light. The light energy is absorbed by green plants and converted into chemical energy stored in the bonds of the organic chemicals formed. For instance, the production of 180 g of glucose (1 mole of glucose) requires an energy input of 2880 kJ at 25 °C and 1 atmosphere pressure. This energy can be released by breaking down the glucose by reaction with dioxygen (Eqn 2.13). The process is called **aerobic respiration**.

$$C_6H_{12}O_6 + 6O_2 \longrightarrow 6CO_2 + 6H_2O + \text{energy} \qquad (2.13)$$
$$\text{glucose}$$

The light emitted from the Sun forms the visible part of the electromagnetic radiation spectrum (Figure 2.2). Electromagnetic radiation can be considered to consist of waves of electromagnetic force whose energy is proportional to their frequency (number of vibrations in a fixed time) and inversely proportional to their wavelength (distance between corresponding parts of adjoining

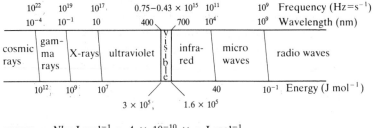

energy $= Nh\nu$ J mol$^{-1} \simeq 4 \times 10^{-10} \times \nu$ J mol^{-1}

frequency $= \dfrac{3 \times 10^{17}}{\text{wavelength (nm)}}$ Hz

Figure 2.2 The electromagnetic spectrum, showing frequencies, wavelengths and energy ranges for different radiation types.

waves, e.g. crest to crest). The relationships between energy and wave properties are expressed by Equations 2.14 and 2.15.

$$\text{energy, } E = Nh\nu \text{ J mol}^{-1} \tag{2.14}$$

or

$$E = Nh\frac{c}{\lambda}\text{J mol}^{-1} \tag{2.15}$$

where N = Avogadro's number = 6.023×10^{23} mol^{-1};
h = Planck's constant = 6.626×10^{-34} J s^{-1};
ν (nu) = frequency, in cycles per second (s^{-1}, or hertz, Hz);
c = velocity of light in a vacuum = 2.998×10^{8} m s^{-1};
λ (lambda) = wavelength, in m (or often nm = nanometre = 10^{-9} m).

Visible light has a wavelength range of about 400 nm (violet light) to 750 nm (red light); this is a frequency range of about 8×10^{14} to 4×10^{14} Hz. By substituting these values in Equations 2.14 or 2.15 we obtain values for the energy of visible light of about 320 kJ mol^{-1} (violet) to 160 kJ mol^{-1} (red).

Green plants contain chlorophyll molecules which are capable of absorbing some of the visible part of the electromagnetic radiation emitted by the Sun. The energy in this absorbed radiation raises the energy of some electrons in the chlorophyll molecules (Figure 2.3). These excited electrons are donated to an electron acceptor that initiates a series of electron-transfer reactions leading eventually to the formation of carbohydrates, such as glucose. The light energy is absorbed at two stages in the photosynthesis system and raises the energy of electrons. The electrons cannot be continuously removed from the chlorophyll molecules without being replaced. In plants

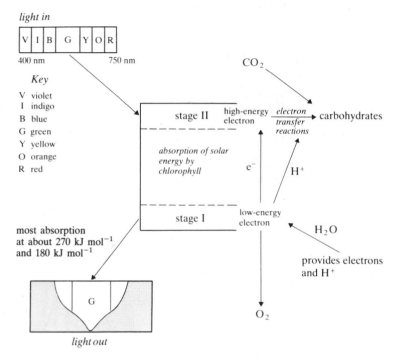

Figure 2.3 The absorption of visible light by chlorophyll during photosynthesis.

and blue-green algae, water provides the replacement electrons (Eqn 2.16), dioxygen is released and the protons eventually combine with carbon.

$$2H_2O \rightleftharpoons 2H^+ + 2OH^- \longrightarrow 4H^+ + O_2 + 4e^- \qquad (2.16)$$

Chlorophyll absorbs mainly red and blue light; therefore the light reflected contains mainly the unabsorbed green light – this is why leaves containing chlorophyll appear green.

Stratospheric ozone and the ozone layer

The electromagnetic radiation from the Sun that falls on the upper layers of the atmosphere contains much more ultraviolet light (i.e. higher-energy radiation) than does the radiation reaching the land and water surface of the Earth (Figure 2.4). This ultraviolet part of the electromagnetic spectrum is often subdivided into three regions of different energy (wavelength, λ). These regions are called UV-C

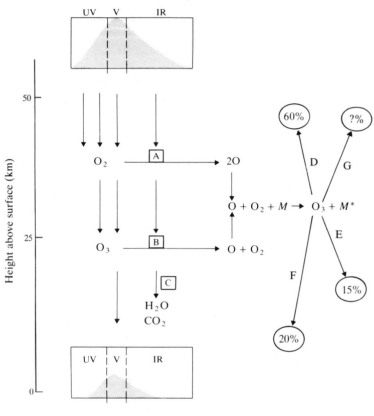

solar radiation above atmosphere

Height above surface (km)

Key

A ultraviolet absorbed, < 242 nm

B ultraviolet absorbed, 200–325 nm

C infrared absorbed

D (i) $NO + O_3 \rightarrow NO_2 + O_2$
(ii) $NO_2 + O \rightarrow NO + O_2$

E (i) $HO + O_3 \rightarrow HO_2 + O_2$
(ii) $HO_2 + O \rightarrow OH + O_2$

F $O + O_3 \rightarrow 2O_2$

G (i) $Cl + O_3 \rightarrow ClO + O_2$
(ii) $ClO + O \rightarrow Cl + O_2$

$n\%$ percentage of natural removal rate

M a molecular species

Figure 2.4 The absorption of ultraviolet radiation by ozone in the Earth's atmosphere. Natural ozone-removal mechanisms are also shown. The ozone layer is at a height of 20–40 km, depending upon latitude.

($\lambda < 290$ nm; highest energy), UV-B ($\lambda = 290$–320 nm; intermediate energy) and UV-A ($\lambda = 320$–400 nm; lowest energy).

The bonds in the dioxygen molecule may be broken (Figure 2.4, reaction A) by UV-C to form free oxygen atoms. At altitudes above 400 km the majority of the oxygen is atomic. At lower altitudes, as the number of dioxygen and dinitrogen molecules increases, ozone, O_3, is formed (Figure 2.4). The molecular species, M, (usually N_2 or O_2) removes the excess energy produced when ozone is formed and so prevents the ozone decomposing immediately. As reaction B (Figure 2.4) indicates, ozone is destroyed by the absorption of the lower-energy end of the UV-C band plus all of the wavelengths in the UV-B band. However, the equilibrium concentration of ozone would be much higher than it actually is if the only reactions involved in its destruction were initiated by ultraviolet radiation and by atomic oxygen (Figure 2.4, reaction F). Some other chemical species in the stratosphere are also able to break down ozone by a reaction sequence that regenerates these species so that they can react again, i.e.

$$O_3 + X \longrightarrow O_2 + XO \qquad (2.17)$$

$$XO + O \longrightarrow O_2 + X \qquad (2.18)$$

The most common naturally occurring species represented by X in reactions 2.17 and 2.18 are NO (nitric oxide) and HO (hydroxyl radical), but recently there have been significant increases in the concentrations of Cl (chlorine) and Br (bromine) in the stratosphere. Fortunately the reaction sequence 2.17 plus 2.18 can be broken by other chemicals reacting with 'X' or 'XO'. Even so, it has been estimated that one atom of chlorine can destroy 100 000 molecules of ozone. The bromine cycle is even less easily broken. NO and HO generally destroy fewer ozone molecules before their cycles are interrupted.

The concentration of ozone in the stratosphere varies with altitude. It reaches a maximum of about ten parts per million by volume (10 ppmv) in the lower stratosphere, above the boundary between the stratosphere and the troposphere. This is the ozone layer.

Most ozone is formed above equatorial regions where the incidence of UV-C is greatest. Once formed the ozone moves towards the poles and also to lower altitudes. There is generally a higher concentration of ozone above polar regions because the rate of removal is slower (less UV-B) than near the equator. The absorption of ultraviolet and infrared radiation (see Chapter 4) in the stratosphere means that the temperature in the stratosphere is higher than at the top of the troposphere. This temperature inversion stabilizes air movements in the stratosphere and controls weather patterns in the troposphere. A reduction in the quantity of ozone in the stratosphere will change the energy absorption properties and may have knock-on effects on the climate at the Earth's surface.

The concentration of ozone in the stratosphere shows daily, seasonal and annual variations of several per cent. Therefore, it is difficult to identify changes in concentration caused by anthropogenic effects unless these are large. Very large reductions have been identified in the spring above the Antarctic – the so-called 'ozone hole'. Whilst the term 'ozone hole' is attention grabbing, it is a misnomer. The term implies an absence of ozone whereas the quantity has rarely been reduced by more than 50%. Measurements of ozone levels have been made at Halley Bay, in Antarctica, since the 1950s. These measurements show that values for October each year were above 300 **Dobson Units** (DU) up to the early 1960s. There were then increasingly wide variations and a general decline, so that by 1980 minimum values had decreased to 230 DU. By 1985 values of 150 DU were recorded in October and since then similar low values have reoccurred. It is clear that:

(a) the lowest values always occur in the spring;
(b) at some altitudes (15–20 km) over 90% of the ozone may be removed;
(c) ozone abundance values of over 350 DU reoccurred in the early summer;
(d) year-on-year value patterns could differ quite considerably; e.g. for Palmer Station, values in 1988 did not fall below 200 DU, whereas in 1989 the lowest value was 170 DU, but recovery to 350 DU occurred 12 days earlier than in 1988.

The detailed explanation of these observations is complex, but the outline features are relatively clear. First, the pattern of events regularly found over the Antarctic is unique, though some similar events occur irregularly over the Arctic. The important requirements for extreme ozone depletion are:

(a) the very low temperatures developed in the atmosphere during the Antarctic winter;
(b) the formation of a stable stratospheric circulation pattern called the Antarctic polar vortex (Figure 2.5);
(c) the increased concentration of chlorine-containing species trapped inside this polar vortex.

The stratospheric air above the Antarctic contains very little water vapour (4–6 ppmv); as a consequence nucleation and condensation

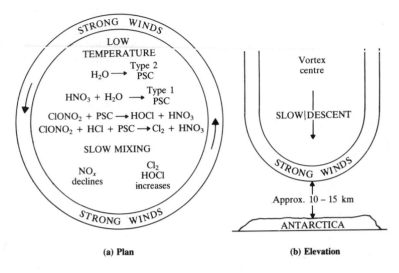

(a) Plan   (b) Elevation

Figure 2.5 Diagrammatic representation of the Antarctic polar vortex system that exists in the winter and early spring. (a) Plan view of the winds circulating above the Antarctic continent and indicating some of the reactions that occur on the surfaces of the polar stratospheric clouds, PSCs. (b) Elevation showing vortex occurs mainly in the stratosphere. The zone of strong winds virtually isolates the inner zone from the rest of the atmosphere.

to form clouds only occurs at very low temperatures. When the temperature drops below 198 K ($-75\,^{\circ}$C) nitric acid trihydrate (NAT) particles, $HNO_3 \cdot 3H_2O$, start to condense to form Type 1 polar stratospheric clouds (Type 1 PSCs). These clouds consist of small particles, similar to ice crystals, with an average diameter of 10^{-6} m and a high surface area. Their formation removes nitrogen oxide species from the gas phase and traps them in the clouds as solid nitric acid. This means that nitrogen dioxide, which can break the ozone destruction cycle initiated by chlorine by reacting with the chlorine monoxide,

$$ClO + NO_2 \longrightarrow ClONO_2 \qquad (2.19)$$

is removed when the nitrogen species become locked up in the clouds. To make things worse, any chlorine nitrate, $ClONO_2$, that is already present in the stratosphere can be destroyed by reactions that occur on the surface of the solid particles, but do not occur in the gas phase, e.g.

$$ClONO_2 + H_2O \longrightarrow HOCl + HNO_3 \qquad (2.20)$$

$$ClONO_2 + HCl \longrightarrow Cl_2 + HNO_3 \qquad (2.21)$$

These reactions increase the quantity of the potentially reactive chlorine species Cl (from Cl_2) and HOCl. The temperatures in the Arctic are not normally as low as in the Antarctic. Therefore, polar stratospheric clouds are formed less frequently and are shorter lived.

The circulation of air currents that arises because of temperature differences between the polar regions and lower latitudes, plus the effect of the Earth's rotation, leads to the formation of a mass of air above the Antarctic that is essentially isolated from the rest of the stratosphere (Figure 2.5). The Antarctic land and ice mass is surrounded by oceans and this favours the setting up of a relatively stable vortex system that can last for several months before breaking up in the spring. The air circulation pattern over the Arctic region is disturbed by the effects of the land masses that radiate out from the North Pole. Therefore, the Arctic vortex system tends to be unstable and rarely lasts for more than a few days. There is little mixing of the air inside the vortex with that outside. The polar stratospheric clouds trapped within the vortex remove nitrogen oxide species from this air and increase the quantities of potentially reactive chlorine species such as HOCl, $ClNO_2$ and Cl_2. With the arrival of spring there is an increase in ultraviolet radiation reaching the stratosphere. The chlorine species are broken down to release free chlorine radicals which then rapidly destroy the ozone in the vortex system by the reaction sequence G in Figure 2.4. As the vortex breaks up and fresh air enters the region the decomposition cycle is quenched (Eqn 2.19) by the new nitrogen oxides that have been brought in. At the same time the ozone is gradually replaced. Before it is finally filled in, the zone of ozone deficiency may move well away from the Antarctic and pass over Australia or South America.

The increased use of chlorine compounds as refrigerants, foaming agents, solvents and in aerosol-spray cans has produced a rapid rise in the concentration of chlorine in the atmosphere. Natural levels of chlorine are about 0.6 parts per billion by volume (ppbv). In 1992 concentrations had reached 3.5 ppbv and are expected to rise to well over 4 ppbv before a decline might occur. The halting of the present rise and the size of the peak value depend upon how effectively emissions of the chlorine compounds are controlled.

The chlorofluorocarbons (CFCs) such as CCl_3F (CFC-11) and CCl_2F_2 (CFC-12) were developed as inert, non-toxic compounds that could be safely used by industry and in the home. Unfortunately they are so inert that when they escape into the atmosphere they slowly pass unchanged through the troposphere and enter the stratosphere. Here UV-C causes photochemical decomposition, yielding free chlorine atoms that catalyse the decomposition of ozone, e.g.

$$CCl_2F_2 \xrightarrow{\text{UV light}} Cl \xrightarrow{O_3} O_2 + ClO \xrightarrow{O} O_2 + Cl$$

dichloro-   activated   chlorine   can react
difluoromethane chlorine   oxide   again
                 atom

(2.22)

The CFCs have long atmospheric lifetimes (65 years for CFC-11 and 130 years for CFC-12). There are already enough of these compounds present in the atmosphere to continue to give elevated stratospheric chlorine concentrations until 2100.

The compounds that have been proposed to replace the CFCs immediately are the hydrochlorofluorocarbons (HCFCs) containing hydrogen atoms, which makes them more likely to break down in the troposphere. Therefore, only a small proportion of the emitted compounds should reach the stratosphere. Compounds being manufactured include CHF_2Cl (HCFC-22) and CCl_2FCH_3 (HCFC-141b). The ultimate goal is fully to replace these chlorinated compounds with chemicals that do not release any ozone-destroying species.

The phasing out of the CFC-type compounds is difficult because replacements are either poisonous (ammonia) or more expensive (HCFCs) which causes problems to poorer countries such as India and China. These countries want to provide refrigerators to most of their people but the higher costs would hold back this programme.

The Halons are similar to the CFCs but contain bromine, e.g. Halon-1301, $CBrF_3$, and Halon-1211, $CBrClF_2$. They are widely used in fire extinguishers, especially in areas where water cannot be safely used. Finding suitable alternatives for these bromine-containing compounds may well be even more difficult than finding alternatives for CFCs.

There is increasing evidence that there are general reductions in ozone concentrations in the stratosphere. The average decrease appears to be only a few per cent in 1992 but it was of variable extent with an especially great reduction in Antarctica. The high natural variability in ozone concentration makes confirmation of these reductions and the identification of harmful effects difficult. The position is further complicated by natural perturbations introduced by large volcanic eruptions. If there is a large amount of sulphate aerosol injected into the stratosphere, these particles can behave like the polar stratospheric clouds and increase the relative availability of ozone-destroying species such as chlorine. The rates of ozone removal in 1992 are thought to have been enhanced by volcanic eruptions in 1991 and 1992. Because anthropogenically introduced chlorine and bromine levels will remain high for so long, it is expected that there will be more and more UV-B reaching the Earth's surface. Before ozone depletion began, 70–80% of the UV-B was absorbed before reaching the Earth's surface. A 10% reduction in stratospheric ozone concentration would reduce the amount of UV-B being absorbed to 55–65%. This could have major effects on susceptible organisms such as plankton and land plants. Effects on humans would include increased risk of skin cancer and eye cataracts. These direct effects on humans are less worrying than the interference with food production that could occur.

The increased use of fertilizers is causing an increase (0.25% per year) in the quantities of dinitrogen oxide, N_2O, being produced by micro-organisms (see Chapter 5). N_2O, like CFCs, is very stable and passes up through the troposphere and enters the stratosphere. In the stratosphere it is broken down under the influence of ultraviolet radiation into a mixture containing about 95% dinitrogen and 5% NO, one of the ozone-destroying compounds (Figure 2.4). The effects of increased N_2O concentrations are at present thought to be relatively small and may well be balanced by increases in methane, CH_4, which should increase ozone concentrations by reacting with chlorine in the stratosphere.

There is still a lot to be learnt about stratospheric chemistry and the balance between production and destruction of ozone. The long lifetimes of many of the chemicals that have been released into the atmosphere are a major source of concern as remedial action will take a long time to become effective.

Tropospheric ozone

The concentration of ozone in the troposphere is very much lower than in the stratosphere. This is just as well, because ozone is toxic to plants (as little as 80 ppbv can reduce growth) and to humans (1 ppmv is fatal and 120 ppbv can cause breathing difficulties). The natural background concentration is in the range 20 to 60 ppbv. However, there is some evidence that in the nineteenth century concentrations were only 10 ppbv.

Though a very small part of the tropospheric ozone is brought down from the stratosphere, most is produced near ground level. This ozone is formed by the action of UV-A ($\lambda < 400$ nm) on nitrogen dioxide:

$$NO_2 \longrightarrow NO + O \qquad (2.23)$$

$$O + O_2 + M \longrightarrow O_3 + M \qquad (2.24)$$

(M is a third body, usually N_2 or O_2, that removes energy released in the reaction.)

Because ozone is also destroyed by nitric oxide the build-up of ozone is dependent on a large number of linked reactions. Some of these will be discussed later in Chapter 5. The important features are that ozone production is favoured by the presence of nitrogen oxides from combustion processes, e.g. petrol engines, industrial furnaces, forest fires, together with hydrocarbons and sunlight. The hydrocarbons may be introduced into the atmosphere from unburnt fuel or may be there naturally (e.g. volcanic or agricultural emissions of methane or emissions of terpenes and isoprene from trees and other plants). Ozone levels in many industrial countries appear to be steadily rising and concentrations greater than 120 ppbv are becoming increasingly common, especially in the spring and summer.

These high tropospheric ozone concentrations are being linked to increasing levels of asthma and bronchial problems amongst urban populations. Similarly, the decline in forest growth that was initially blamed on 'acid rain' is now recognized as being partially due to the detrimental effects of ozone. Human activities that reduce stratospheric ozone concentrations and thus increase UV-B radiation levels are helping to increase ground-level ozone concentrations. The more UV-B that reaches the lower levels of the troposphere, the more likely that harmful ozone will be produced in that region of the atmosphere where it can do maximum damage to crops and humans.

Part Two

Major elements found in living matter

There are 90 naturally occurring elements on Earth and, provided sensitive-enough analytical methods are used, all 90 elements are likely to be detected in living organisms. However, in humans 99.3% of all atoms are carbon, hydrogen, nitrogen and oxygen. Other organisms have a similar overwhelming preponderance of these four elements. We shall study the chemistry of hydrogen, carbon and nitrogen in this section of the book. In addition, the chemistry of sulphur and phosphorus will be considered. Sulphur is an important minor component of many of the molecules found in organisms. Phosphorus is also found in some biologically important molecules, but the chemical changes that it undergoes in natural systems are relatively limited.

Living organisms are characterized by the presence of cells – 'the smallest unit of life' (Figure 1.4). In cells occur the many reactions that are necessary for the energy production and continuing maintenance of the organism. The cell has an outer membrane which acts as a container for the internal components and also regulates the passage of chemicals into and out of the cell. Cells consist of about 70% water, by weight, and 28.5% organic molecules. Most of the wide variety of organic compounds making up the cells can be arranged into four groups. These groups are (a) carbohydrates, (b) lipids, (c) nucleic acids, and (d) proteins.

As Figure P2.1 illustrates, the organic molecules found in cells are actually produced by the linking together of smaller molecules. In each case, when two of the smaller molecules combine a molecule of water is released. This is called a **condensation reaction**. The reverse reaction, which breaks down the large molecules into their constituent smaller molecules, occurs in the presence of water, usually requires a **catalyst**, and is called **hydrolysis**. If a large number of small

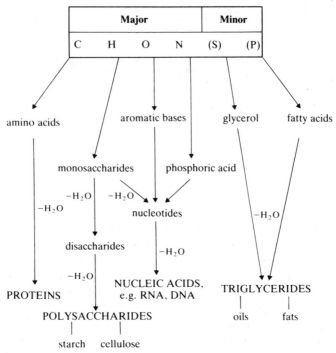

Figure P2.1   Inter-relationships between some organic molecules and their constituent elements.

molecules are used, as in the production of starch, cellulose, proteins and nucleic acids, the large molecules are called **polymers** and the small molecules are **monomers**.

Carbohydrates

Carbohydrates generally have the common formula $C_x(H_2O)_y$. The simplest carbohydrates are the monosaccharides which usually contain either five carbon atoms, e.g. ribose, $C_5H_{10}O_5$, or six carbon atoms, e.g. glucose and fructose, $C_6H_{12}O_6$. These compounds consist of either five- or six-membered rings containing one oxygen atom and four or five carbon atoms respectively. These simple units can be cross-linked to form disaccharides (two units), e.g. sucrose (1 glucose + 1 fructose: $C_{12}H_{22}O_{11}$); or polysaccharides (up to several thousand units), e.g. starch and cellulose. Starch is used as a reserve foodstuff in plants and to a lesser extent in animals. Cellulose acts as a supporting structure for the cells of plants and is the most abundant type of organic compound on the Earth. Unfortunately, humans cannot digest cellulose so it cannot be directly used as a food.

Lipids

Lipids are a rather diverse group of compounds that are soluble in organic (non-polar) solvents – such as trichloromethane (chloroform, $CHCl_3$), benzene (C_6H_6) and ethoxyethane (ether, $C_2H_5OC_2H_5$) – but insoluble in water. If the lipid is solid at room temperature it is called a **fat** (most animal-derived lipids), and if it is liquid it is called an **oil** (most vegetable-derived lipids). The majority of the commonly found lipids are formed by the combination of glycerol (an alcohol with three hydroxyl, OH, groups) and fatty acids (hydrocarbons with a carboxylic acid, – COOH, group; Figure P2.2). The letter 'R' in a

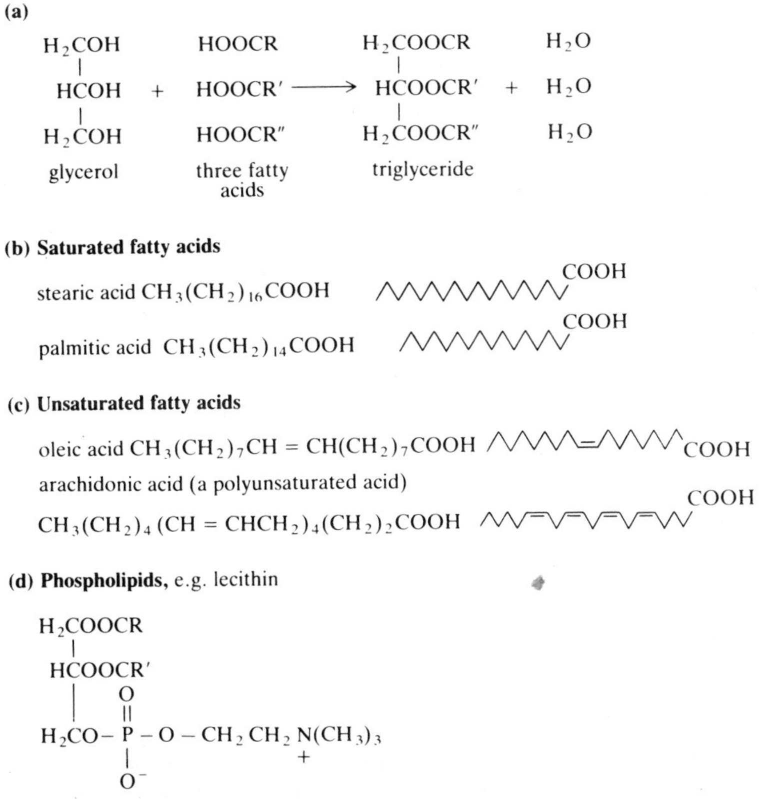

(a)

$$
\begin{array}{cccc}
H_2COH & HOOCR & H_2COOCR & H_2O \\
| & & | & \\
HCOH & + \quad HOOCR' \longrightarrow & HCOOCR' & + \quad H_2O \\
| & & | & \\
H_2COH & HOOCR'' & H_2COOCR'' & H_2O \\
\text{glycerol} & \text{three fatty} & \text{triglyceride} & \\
& \text{acids} & &
\end{array}
$$

(b) **Saturated fatty acids**

    stearic acid $CH_3(CH_2)_{16}COOH$

    palmitic acid $CH_3(CH_2)_{14}COOH$

(c) **Unsaturated fatty acids**

    oleic acid $CH_3(CH_2)_7CH = CH(CH_2)_7COOH$

    arachidonic acid (a polyunsaturated acid)

    $CH_3(CH_2)_4 (CH = CHCH_2)_4(CH_2)_2COOH$

(d) **Phospholipids,** e.g. lecithin

$$
\begin{array}{l}
H_2COOCR \\
| \\
HCOOCR' \\
| \quad\quad O \\
| \quad\quad || \\
H_2CO- P - O - CH_2 CH_2 N(CH_3)_3 \\
| \quad\quad\quad\quad\quad\quad\quad\quad + \\
O^-
\end{array}
$$

Figure P2.2 Lipids. (a) Formation of a triglyceride. Structures of (b) saturated fatty acids, (c) unsaturated fatty acids, and (d) phospholipids.

formula is used to indicate any organic group, usually only containing carbon and hydrogen. If the carbon atoms in the group R are joined to each other by single bonds, both the acid and the lipid it forms are described as being **saturated**. **Unsaturated** acids and fats have double bonds between some of the carbon atoms. The greater the degree of unsaturation, i.e. the larger the number of carbon–carbon double bonds, the lower the melting point of the lipid.

It should be noted that the oil ('mineral oil') that is extracted from the ground and used as the basis of the petrochemical industry is mainly composed of hydrocarbons rather than triglycerides. Chemically mineral oils and lipid oils are quite different.

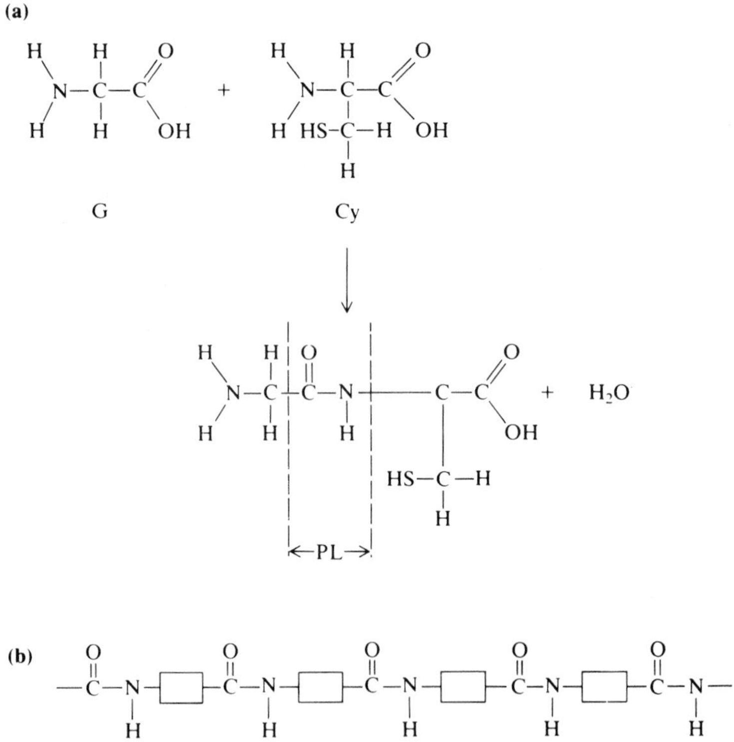

(a)

(b)

Figure P2.3 The formation of peptide linkages between amino acids to produce proteins. (a) Formation of peptide linkage (PL) between glycine (G) and cysteine (Cy), producing a dipeptide. (b) Diagrammatic representation of a protein, a polypeptide, showing peptide linkages between amino acid residues.

In phospholipids, the glycerol is combined with two fatty acids and a substituted phosphoric acid occupies the third position (Figure P2.2). The phosphate group is relatively **polar** (water soluble). Cell walls are composed of double layers of the phospholipids, with the polar group as the exterior of the wall (in contact with aqueous media) and the fatty-acid, **non-polar**, groups making up the interior of the wall.

Nucleic acids

Nucleic acids are polymers composed of an aromatic base (e.g. adenine), plus a pentose sugar (e.g. ribose or 2-deoxyribose) and phosphate groups (PO_4^{3-}) which, together, form a nucleotide unit. The nucleotide units are then linked together to form the nucleic acid, e.g. ribonucleic acid (RNA) or deoxyribonucleic acid (DNA). The nucleic acids store and transmit genetic information.

Proteins

Proteins are polymers of amino acids formed by the combination of the acidic ($-COOH$) group of one amino acid with the basic amino ($-NH_2$) group of another amino acid (Figure P2.3) to form a peptide bond. Proteins are large polypeptides with molecular weights greater than 10 000.

Proteins may be soluble or insoluble in water. The latter group are utilized in animals for both structural and connective tissue. They are also important components of enzymes, which are catalysts.

3 Hydrogen

Hydrogen Abundance by weight (the relative abundance is given in parentheses): Earth 78 ppm (18); crust, 0.22% (10); ocean, 11% (2); atmosphere as H_2, 0.5 ppm (9). ppm = mg kg^{-1}.

Hydrogen is by far the most abundant element in the Universe. Over 90% of all atoms are hydrogen. The next most common element is helium, He, with an abundance of 8 or 9%. All the other atoms taken together constitute less than 1% of the material in the Universe. In contrast, hydrogen is only a minor component of the Earth, though its presence in water on the Earth's surface has been critical in providing the right conditions for the development of life on the planet.

It is thought that all the elements were originally derived from hydrogen, being formed by nuclear reactions occurring in stars. These reactions are still continuing and the original hydrogen is gradually being consumed. Stars form when a gravitational collapse of a hydrogen gas cloud releases energy to produce a concentration of hotter, denser material. When the temperature rises to about 10 million degrees Kelvin, the hydrogen nuclei can combine to form helium (Eqn 3.1).

$$2{}_1^1\text{H} \xrightarrow{\hspace{1cm}} {}_1^2\text{H} \xrightarrow{{}_1^1\text{H}} {}_2^3\text{He} \xrightarrow{{}_2^3\text{He}} {}_2^4\text{He} + 2{}_1^1\text{H} \quad (3.1)$$
$$\text{hydrogen} \qquad\qquad \text{deuterium} \qquad\qquad \text{helium-3} \qquad \text{helium-4}$$

The combination of two lighter nuclei to form a heavier nucleus is called **nuclear fusion**, and this series of nuclear-fusion reactions liberates 2.5×10^{12} J for each 4 g of ${}_2^4\text{He}$ produced. (As a comparison 1 gallon of petroleum – about 4 kg – liberates 1.2×10^8 J when burnt.) When the concentration of ${}_2^4\text{He}$ becomes high enough, further nuclear reactions can occur leading to the formation of carbon, ${}_6^{12}\text{C}$ (Eqn 3.2).

$$2{}_2^4\text{He} \xrightarrow{\hspace{1cm}} {}_4^8\text{Be} \xrightarrow{{}_2^4\text{He}} {}_6^{12}\text{C} \qquad\qquad (3.2)$$
$$\text{beryllium} \qquad\qquad \text{carbon}$$

Once ${}_6^{12}\text{C}$ has been produced a number of reactions can occur that involve the conversion of more ${}_1^1\text{H}$ to ${}_2^4\text{He}$ and the addition of ${}_2^4\text{He}$ to larger nuclei to produce oxygen (${}_8^{16}\text{O}$), neon (${}_{10}^{20}\text{Ne}$), magnesium

($^{24}_{12}$Mg), silicon ($^{28}_{14}$Si) and so on up to iron ($^{56}_{26}$Fe). The $^{56}_{26}$Fe nucleus is the most stable of all the nuclei, at temperatures below 30×10^9 K. Heavier nuclei, and those with odd numbers of protons, are mainly produced in stars by neutron-capture reactions (Eqn 3.3), the same process that is used to produce isotopes commercially in nuclear reactors on Earth.

$$\underset{\text{iron-56}}{^{56}_{26}\text{Fe}} + \underset{\text{neutrons}}{3^{1}_{0}\text{n}} \longrightarrow \underset{\text{cobalt-59}}{^{59}_{27}\text{Co}} + \underset{\text{electron}}{^{0}_{-1}\text{e}} \tag{3.3}$$

Because neutrons have no charge they can penetrate the nucleus of atoms relatively easily and may induce instability that leads to the formation of new elements.

The stability of a nucleus depends upon the balance between the attractive forces and the repulsive forces. The major repulsive force is due to the electrostatic interaction between the positive charges of the protons. The attractive forces are of very short range and depend upon the total number of **nucleons**, i.e. protons plus neutrons. As the number of protons increases, the repulsive forces increase more rapidly than the attractive forces unless extra neutrons are added to counterbalance this effect. With the lighter elements the neutron:proton ratio for stable nuclei is about 1:1 – e.g. $^{4}_{2}$He (2p + 2n), $^{16}_{8}$O (8p + 8n) – but with the heavier nuclei the proportion of neutrons required for stability becomes greater – e.g. nickel, $^{58}_{28}$Ni (28p + 30n), lead, $^{206}_{82}$Pb (82p + 124n) – as shown in Figure 3.1. Many elements have more than one stable isotope. For instance, oxygen has three stable isotopes, $^{16}_{8}$O, $^{17}_{8}$O and $^{18}_{8}$O. Tin has ten stable isotopes – the largest number of any element.

Unstable nuclei become stable when the neutron:proton ratio is adjusted so that the attractive forces become stronger than the repulsive forces. The unstable nuclei are said to be **radioactive**. There are four mechanisms by which unstable nuclei achieve stability (Figure 3.1).

Neutrons in excess: β^- (beta minus) emission

$$\underset{\text{neutron}}{^{1}_{0}\text{n}} \longrightarrow \underset{\text{proton}}{^{1}_{1}\text{p}} + \underset{\text{electron}}{\beta^-(^{0}_{-1}\text{e})} + \underset{\text{gamma rays}}{\gamma} \tag{3.4}$$

One of the neutrons is converted to a proton, and an electron is ejected from the nucleus. There is an increase in atomic number and a new element is produced (Eqn 3.5).

$$\underset{\text{hydrogen-3}}{^{3}_{1}\text{H}} \longrightarrow \underset{\text{helium}}{^{3}_{2}\text{He}} + ^{0}_{-1}\text{e} \tag{3.5}$$

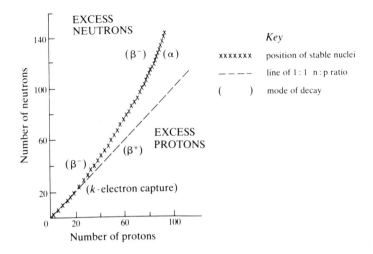

Figure 3.1 The neutron:proton ratio of nuclides, and its relationship to nuclear stability and radioactive-decay mechanisms.

Further lowering of energy is achieved by the emission of **gamma radiation** or **neutrinos**. Gamma rays are very short-wavelength, high-frequency, electromagnetic radiation (Figure 2.2), and neutrinos are small neutral particles. When radioactivity was first discovered, various types of radiation were recognized and named α (alpha), β (beta) and γ (gamma) after the first three letters of the Greek alphabet. The names are still used today even though the constitution of the radiation is known.

Protons in excess: β^+ (beta-plus or positron) emission

$$\underset{\text{proton}}{{}^1_1\text{p}} \longrightarrow \underset{\text{neutron}}{{}^1_0\text{n}} + \underset{\text{positron}}{\beta^+({}^0_1\text{e})} \tag{3.6}$$

A **positron** has the same properties as an electron except that it has a positive charge rather than a negative charge. The emitted positron collides with an electron and they are both destroyed, giving out gamma radiation. Positron emission is often found with artificially produced radio-isotopes. The atomic number decreases and a new element is formed (Eqn 3.7).

$$\underset{\text{copper}}{{}^{64}_{29}\text{Cu}} \longrightarrow \underset{\text{nickel}}{{}^{64}_{28}\text{Ni}} + \underset{\text{positron}}{{}^0_1\text{e}} \tag{3.7}$$

Protons in excess: α (alpha) emission

An alpha particle, which is a helium-4 nucleus ($^4_2He^{2+}$), is emitted from heavy unstable nuclei causing the loss of two neutrons and two protons. This type of decay is only effective for elements that have more neutrons than protons, i.e. lie above the 1:1 line (Figure 3.1) so that the proportion of protons is reduced (Eqn 3.8).

$$\underset{\text{uranium}}{^{238}_{92}U} \longrightarrow \underset{\text{thorium}}{^{234}_{90}Th} + \underset{\text{alpha particle}}{^{4}_{2}He} \tag{3.8}$$

$$\left(\frac{n}{p} = \frac{146}{92} = 1.59\right) \qquad \left(\frac{n}{p} = \frac{144}{90} = 1.60\right)$$

Protons in excess: electron capture

The nucleus captures an electron from the lowest-energy group of electrons (the 1s or 'K shell'), converting a proton into a neutron (Eqn 3.9). Another, higher-energy, electron drops into the 'K shell', and X-rays, short-wavelength electromagnetic radiation (Figure 2.2), are emitted.

$$\underset{\text{proton}}{^{1}_{1}p} + \underset{\text{electron}}{^{0}_{-1}e} \longrightarrow \underset{\text{neutron}}{^{1}_{0}n} \tag{3.9}$$

Again a new element is produced (Eqn 3.10).

$$\underset{\text{potassium}}{^{40}_{19}K} \xrightarrow[\substack{^{0}_{-1}e}]{\text{K electron capture}} \underset{\text{argon}}{^{40}_{18}Ar} + \underset{\substack{\text{gamma}\\\text{ray}}}{\gamma} + \underset{\text{X-rays}}{x} \tag{3.10}$$

3.1 Isotopes of hydrogen

Naturally occurring hydrogen, on the Earth, consists of 99.985% ^1_1H, 0.015% ^2_1H, and the radioactive ^3_1H, whose concentration in surface ocean water has been estimated as $1 \times 10^{-18}\%$. Hydrogen is the only element whose isotopes are given specific names – ^1_1H, protium (symbol P); ^2_1H, deuterium (symbol D); ^3_1H, tritium (T).

As we have already seen, nuclear-fusion reactions can liberate very large amounts of energy. The uncontrolled release of this energy occurs when hydrogen bombs are exploded. Nuclear-fusion reactions only occur at very high temperatures ($5-100 \times 10^6$ K), so a very large amount of energy must be put into the system before the greater amount of energy that nuclear-fusion reactions produce can be released. The problems are how to achieve and maintain the initiating temperature and then how to extract the energy in a controlled manner from this hot mass. In the hydrogen bomb an explosion is used to raise the temperature to the required value: no attempt is made to control the subsequent release of energy from the fusion reaction. The advantage of nuclear-fusion reactions as an energy source is that, if deuterium can be utilized as the fuel, existing fuel reserves should last for at least 1 million years, as compared with the 400 years that fossil fuels (coal, oil and natural gas) are forecast to last. Two of the possible nuclear-fusion reactions, all of which involve isotopes of hydrogen, are given in Equations 3.11 and 3.12.

$$2\,^2_1\text{H} \longrightarrow \,^3_2\text{He} + \,^1_0\text{n} \qquad (3.11)$$

$$2\,^2_1\text{H} \longrightarrow \,^3_1\text{H} + \,^1_1\text{H} \qquad (3.12)$$

The environmental problems associated with nuclear-fusion reactors, if they are ever built, appear to be less than the problems of present-day nuclear-fission reactors, but until we are much closer to seeing how the reactor will actually operate the problems cannot be completely identified. **Nuclear fission** involves the breaking up of large nuclei, such as ^{235}U, by bombarding them with neutrons. This break-up releases less energy than nuclear-fusion reactions, but more energy than chemical reactions, for a given mass of fuel. Unfortunately the initial products of nuclear-fission reactions are radioactive, and the disposal of this dangerous waste presents a major problem.

Tritium is produced naturally by the interaction of cosmic rays (protons, electrons, various nuclei and nuclear particles coming from outer space) with gas in the upper atmosphere. The tritium gradually changes into stable ^3_2He by a radioactive-decay process (Eqn 3.5) with β^- emission. The decay reactions are governed by statistical laws and the decay is a first-order reaction (Eqn 3.13). This means that the number of disintegrations in a fixed time period from a group of radioactive atoms is proportional to the number of radioactive atoms present (Eqn 3.14). As the disintegrations occur, the number of radioactive atoms remaining becomes smaller and thus the number of disintegrations in the given time period also decreases – a typical exponential relationship.

$$\text{rate of disintegration,} \quad \frac{dN}{dt} = -\lambda N \qquad (3.13)$$

$$N = N_0 e^{-\lambda t} \qquad (3.14)$$

where N = number of atoms present at time t;

λ = decay constant, i.e. the probability of a nucleus disintegrating in unit time;

N_0 = number of atoms originally present, i.e. when $t = 0$.

A convenient expression for comparing rates of decay is the **half-life**, $t_{\frac{1}{2}}$, i.e. the time taken for an initial number of atoms (N_0) to be reduced to half that number: $N = \frac{1}{2}N_0$ when $t = t_{\frac{1}{2}}$. The half-life for tritium is 12.5 years. If we start with a fixed number of tritium nuclei, say 2000, then after 12.5 years there will be 1000 nuclei, after 25 years 500 nuclei, and after 37.5 years 250 nuclei. In practice, because the decay process is random, there would not be exactly 1000, 500, 250, etc., but 95.4% probability that the value would be 1000, 500, etc., plus or minus two times the standard deviation. The half-life of tritium is so short compared to the age of the Earth that the only way that there can be any tritium present now is for the supply to be continuously replenished. (Even if the whole mass of the Earth had been tritium, i.e. contained about 10^{51} atoms of ^3_1H, after 169.4 half-lives, or 2117 years, there would be no atoms of ^3_1H left.) In addition to natural sources, tritium is now produced by nuclear explosions, in nuclear reactors and in nuclear-fuel reprocessing plants. Because of tritium's short half-life the stopping of these military and industrial activities would allow the reduction back to natural levels to occur relatively quickly. The tritium released, usually as a gas, exchanges with hydrogen in water and the tritiated water can enter living organisms as part of the ingested water. When a tritium nucleus decays it emits a β^- particle (on average these each have an energy of 3×10^{-15} J). This energy is so low that the electron will not penetrate more than a few millimetres of air and would be stopped, for example, by human skin. However, the tritiated water can enter cells and even low-energy β particles are capable of causing disruption to metabolic processes. β particles are ejected from the atomic nucleus with a high velocity: when they collide with an atom they give up some of their energy to that atom, often causing one of the outer electrons to be lost and so producing a positive ion. The particle continues to collide with atoms, causing further ionization, until it loses its kinetic energy and comes to a stop. The higher the initial energy of the β particle, the more penetrating it is and the greater the amount of ionization it can cause. It is the transfer of energy to molecules in living cells that causes malfunctioning and hence the concern about the health effects of radioactive materials.

3.2 Water

Hydrogen is only a minor component of the Earth as a whole, but water is of major importance to the survival of life on the planet. Water covers about 70% of the Earth's surface and the properties of this liquid and its vapour control the climatic conditions that make life possible on Earth. In addition, water's solvent properties control the chemical weathering of rocks, the transfer of nutrients to plants and the transfer of chemicals inside organisms.

Of all water within the surface zone of the Earth, 97% is in the oceans (Figure 3.2), about 2% is in ice caps and glaciers, which cover 10% of the present land surface, and only 0.6% is fresh water of direct use to humans. The water cycle is driven by the absorption of solar energy, most of which causes evaporation of water from the oceans and land though a small proportion generates the winds, waves and currents that aid the circulation in both the atmosphere and water masses; 86% of the water evaporated comes from the oceans, but only 78% of the rain and snow that falls comes down on the oceans. There is a net transfer of water from the oceans to the land so that the precipitation on to the land is 57% greater than the evaporation from the land. The extra water added to the land eventually returns to the oceans via surface runoff in rivers or direct seepage of ground water into the oceans. The average 'residence time' for water in the atmosphere is 11 days. The evaporation of water requires the absorption of energy and tends to reduce the temperature at the air/water interface.

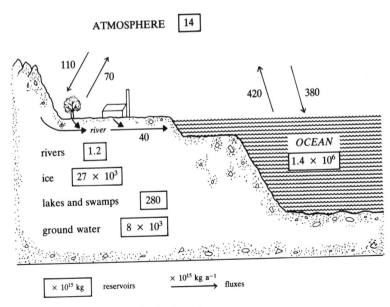

Figure 3.2 The water cycle, in simplified form.

This energy is released again when the water vapour condenses, so the movement of water vapour transfers heat energy from one area to another. As water evaporates most rapidly wherever the temperature is highest and condenses when the temperature drops, the water cycle is effective in reducing the temperature differences between areas. Though the average precipitation over all land masses is equivalent to about 700 mm a year, there are vast differences between different areas both in total amount and in seasonal variations – compare the Atacama Desert, Chile, which has 0 mm a^{-1}, with Cherrapunji, India, which has 26×10^3 mm a^{-1}. In Britain the mean value for lowlands is 500 mm a^{-1}, with 2500 mm a^{-1} in western hill areas.

The water that enters the atmosphere is composed of water that has evaporated from salt- or fresh-water sources leaving the dissolved components behind. However, small sea-spray droplets thrown into the air undergo rapid evaporation, leaving fine salt particles suspended in the atmosphere. The majority of these particles return to the ocean, but sufficient are carried over the land to be washed out by rain, or snow, and to make a significant addition of sodium, Na$^+$, and chloride, Cl$^-$, ions, especially to those areas close to the sea. The rain also dissolves other soluble components present in the atmosphere, such as dioxygen, dinitrogen, carbon dioxide and the oxides of sulphur and nitrogen. In addition dust and other insoluble particulate matter is washed out. In areas where there are large inputs of sulphur oxides and nitrogen oxides, particularly associated with industrial processes and fossil-fuel combustion, the rain may become between ten and a hundred times more acidic than normal.

The rain water that runs over and through the surface of the land dissolves soluble species. The weathering process is speeded up by bringing reactive species, such as acids, into contact with the minerals of the soil and rocks, and by the removal of the reaction products. This results in the water entering rivers having more dissolved solids than the rain. The residence time of the water on land varies from a few days with surface runoff to hundreds or thousands of years for ground water and ice caps. On entering the sea there is a region of mixing of river and sea water, often in an estuary, which is very complex chemically. As well as there being rapid changes in the total concentration of all dissolved solids, individual soluble species may become insoluble and previously insoluble components of the suspended solids carried by the rivers may become soluble as the chemical conditions alter in the zone of mixing. The levels of dissolved species in the oceans appear to be in equilibrium though there are local variations. The residence time of water in the oceans is about 4000 years.

The soluble components in river water are derived partly from atmospherically recycled marine salts deposited on the land by rain or snow and partly from the weathering of rocks and soils. The characteristic cation in sea water is sodium whereas in fresh water it is calcium. If the total dissolved-solids content of natural waters is plotted against the weight ratio of dissolved sodium: dissolved sodium plus calcium (Figure 3.3), a clear relationship between the three factors can be seen. The rivers with the lowest dissolved-solids contents have a high proportion of sodium, indicating that most of the dissolved ions came from precipitation (rain or snow). A comparison (Figure 3.4) of the concentrations of the major dissolved cations in rain water falling on the catchment of the Rio Tefé (1 in Figure 3.3a) and the ions in the river water shows that apart from an increase in potassium and silicon there is little difference. Rivers with these types of dissolved solids are described as being precipitation dominated (Figure 3.3b). They are usually in areas with low relief where weathering has almost reached the end of its cycle and the soils contain only the most stable mineral components.

The water running off the land picks up more soluble species in areas where weathering is still in its earlier stages, so the total dissolved-solids content rises and the proportion of calcium increases. The actual proportion of calcium to sodium will depend upon the rock types being weathered. Catchments with high proportions of limestones (31% Ca) or basic igneous rocks (7% Ca) will obviously have much higher proportions of calcium than catchments draining acid igneous rocks (1.6% Ca) or sandstones (1–2% Ca). (See Part Three.) As rivers flow towards the sea there is some evaporation of the water. In arid areas where evaporation exceeds rainfall, the concentration of dissolved species will be markedly increased as the water evaporates. If the concentration of calcium rises sufficiently for the solubility of calcium carbonate ($6.7 \times 10^{-3}\,\mathrm{g\,dm^{-3}}$) to be exceeded, precipitation of calcium carbonate will occur and the proportion of sodium will correspondingly increase. The ultimate product of this process is sea water. Rivers appearing on this part of the diagram are dominated by evaporation–crystallization processes, and their concentration relationships change along their length, e.g. the Rio Grande (2 in Figure 3.3a). The use of river water for irrigation will increase its dissolved load by passing it through the soil, allowing more soluble components to be picked up and also increasing the time for evaporation to occur. This can be a major problem as the higher the dissolved-species content, the less suitable the water becomes for drinking, irrigation and industrial usage.

Properties of water

All three **phases** of water (solid, liquid and gas) are found on the Earth's surface. The formula for water, H_2O, indicates a molecular weight of 18. If we look at other covalently bonded molecules with similar molecular weights (e.g. CO and NO) we find that they have boiling points below the normal temperature range, say 258 K to 323 K ($-15\,°C$ to $50\,°C$), found on the Earth. The temperature of the melting point indicates the strength of the forces holding the particles together in the solid. Similarly the boiling-point temperature indicates the strength of the forces holding the molecules together in the liquid. Ice and water have relatively strong attractive forces between the molecules.

The water molecule consists of an oxygen atom surrounded by four pairs of electrons (Figure 3.5a). Two of the pairs are shared with hydrogen atoms to form covalent bonds between the oxygen and hydrogens. The other two pairs are called lone pairs because they are not shared with another atom. The pairs of electrons, with their associated negative charges, tend to repel each other and they move as far apart as possible (Figure 3.5b). The maximum separation is achieved if the four pairs are arranged tetrahedrally (Figure 3.5c). However, the lone pairs of electrons occupy a smaller volume and have a greater repulsive effect than the shared pairs of electrons, thus the H–O–H bond angle is reduced from $109.5\,°$ to $104.5\,°$ (Figure 3.5).

Oxygen is more electronegative than hydrogen and attracts the shared bonding pair of electrons more strongly than does the hydrogen. This results in the oxygen having a slight negative charge and each of the hydrogens having a small positive charge (Figure 3.6a). The extra negative charge is concentrated on the two lone pairs of electrons (Figure 3.6b). The water molecule is described as being **dipolar**, as it has one end negative and the other end positive. The negative end of one molecule is electrostatically attracted to the positive end of a neighbouring molecule. Because there are effectively two negative charges, one on each lone pair, each oxygen can be electrostatically attached to two hydrogen atoms from other water molecules (Figure 3.7). This electrostatic bond is called a **hydrogen bond**. The inter-molecular hydrogen bond is about one-tenth as strong as the normal covalent oxygen–hydrogen bond. It is the need to break these hydrogen bonds that is responsible for the high melting and boiling points of water.

Figure 3.3   The chemistry of the Earth's surface waters: (a) variations in proportions of sodium and calcium compared to total dissolved solids; (b) processes controlling the chemistry of the Earth's surface waters. (After Gibbs, R. J. 1970. Mechanisms controlling world water chemistry. *Science* **170**, 1088–90. Copyright © 1970 AAAS.)

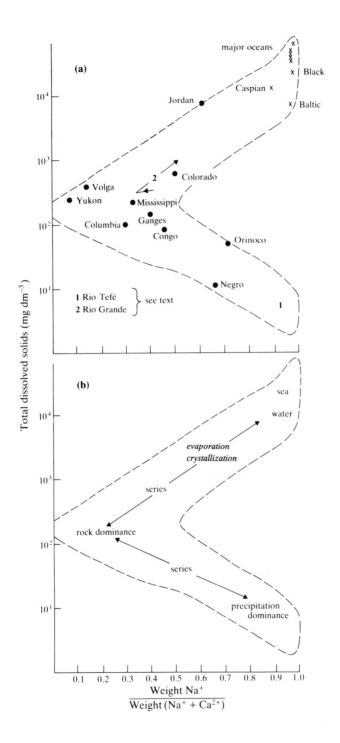

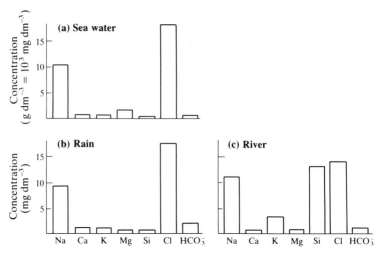

Figure 3.4 Comprison of the major dissolved components in (a) sea water, (b) rain water falling on the Rio Tefě basin, and (c) the Rio Tefé (1 in Figure 3.3a). (After Gibbs, R. J. 1970. Mechanisms controlling world water chemistry. *Science* **170**, 1088–90. Copyright © 1970 AAAS.)

Each water molecule can be involved in four hydrogen bonds: two hydrogen bonds to the oxygen via its lone pairs, and a hydrogen bond between each hydrogen and the oxygen in another water molecule. In ice all the possible hydrogen bonds are formed, and this gives an extended, open, three-dimensional structure.

When ice melts to form water only about 15% of the hydrogen bonds are broken. The open crystalline structure collapses and the volume occupied by the same mass of water molecules is reduced, i.e. the density increases. As more thermal energy is put into the system, (a) the temperature begins to rise, (b) more hydrogen bonds are broken and the structure collapses even further, so increasing the density, and (c) the molecules vibrate more as their kinetic energy increases and so they occupy a greater volume. The changes (b) and (c) produce opposite effects and the result is that fresh water has a density maximum at 3.98 °C. At higher temperatures the vibrational effects are greater than the collapse effect.

Because of these density changes the behaviour of water is particularly suited to the preservation of aquatic life-forms. A lake in summer, if deep enough, becomes stratified (Figure 3.8) due to the poor conductivity of the water, allowing the warming up of the surface layers which become less dense than the cooler lower layers. As the two layers do not mix, the lower layer (hypolimnion) will tend to become oxygen deficient as the organic matter present is oxidized. At the same time

(a) **Four pairs of electrons surrounding the oxygen – two bonding pairs shared with hydrogens plus two lone pairs**

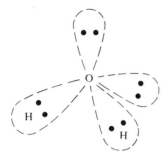

(b) **The repulsive forces between the electron pairs**

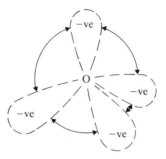

(c) **A tetrahedral arrangement – all angles A, B, C, D are 109.5°**

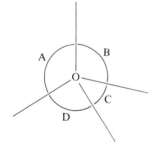

(d) **The H—O—H bond angle, 104.5°, is less than the tetrahedral angle**

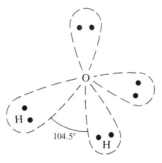

Figure 3.5   Structure of the water molecule.

the upper layer (epilimnion) can become deficient in nutrients, as the living organisms remove the nutrients from the water and then, on dying, fall to the lake bottom, taking the consumed nutrients with them. In the autumn, air temperatures drop and the surface layers of the lake cool down. When the epilimnion temperature drops to that of the hypolimnion, the stratification disappears and vertical mixing, or turnover, occurs (Figure 3.8b). This allows oxygenation of all the water and carries nutrients up from the lower levels to the surface. As the air temperature continues to fall, the surface water temperature drops below 4 °C and the surface layer is again less dense than the bottom layer at 4 °C. When the surface temperature becomes 0 °C,

(a) **Even sharing of bonding electrons between oxygen and hydrogen**

(b) **Oxygen is more electronegative, so electrons spend more time closer to oxygen than hydrogen giving small time-averaged negative, $\delta-$, and positive, $\delta+$, charges**

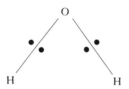

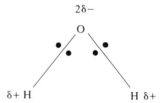

(c) **The small negative charges, $\delta-$, on the oxygen tend to be associated with the lone pairs**

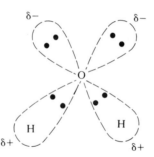

Figure 3.6   The dipole structure of water molecules.

ice will form on the surface, leading to the release of energy as the extra hydrogen bonds are formed: this energy raises the temperature of the water below the ice. Once the ice covers the surface, it acts as an insulating layer, reducing the heat loss from the water under the ice to the air above. The rate of freezing is reduced: the thicker the ice, the lower the rate of freezing. Unless a lake is very shallow or the air temperature remains below 0 °C all year, the water never completely freezes and aquatic life can continue below the surface. In spring, the air temperature rises and the ice that is in contact with the warmer air melts. The water then warms up and again there is the possibility of

(a) **The formation of hydrogen bonds between lone pairs on the oxygen and the hydrogens of neighbouring molecules**

(b) **The four hydrogen bonds from one water molecule – this creates an open structure by holding the water molecules apart in fixed positions**

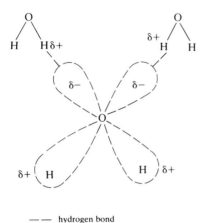

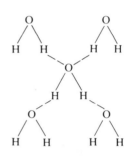

—— hydrogen bond

Figure 3.7 The formation of hydrogen bonds between water molecules.

complete mixing when the surface and bottom temperatures become the same. The stratified structure then gradually develops in the late spring and summer.

Physical weathering of rocks and soils is helped by the expansion of water when it freezes. Water trapped in cracks builds up very large forces when it freezes and attempts to expand. These forces can cause the rocks to shatter in the same way that water pipes burst when the water in them becomes frozen.

When thermal energy is added to a substance, the **kinetic energy** of the component molecules increases. This means that the molecules vibrate and in liquids and gases they also move more rapidly from place to place. The amount of energy needed (a) to break enough hydrogen bonds to convert ice to liquid water is $320 \, \text{J g}^{-1}$; (b) to raise the temperature from the melting point, 273 K, to the boiling point, 373 K, is $420 \, \text{J g}^{-1}$; (c) to break enough hydrogen bonds to convert liquid water to water vapour is $2260 \, \text{J g}^{-1}$. In each case the amount of heat required to bring about the changes (a), (b) and (c) is very much larger than for most substances. Therefore water is a very good medium for controlling temperature changes and transferring heat. In living organisms the chemical reactions occurring in cells often generate a large amount of heat that can be absorbed by the water with only a

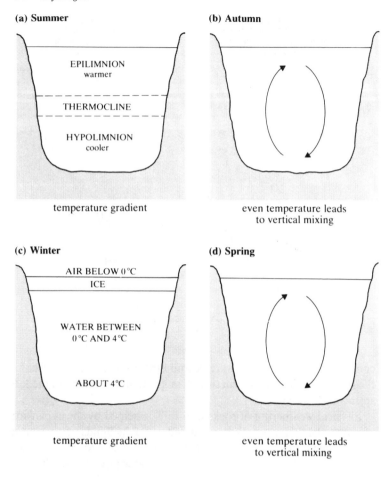

(a) Summer

EPILIMNION
warmer

THERMOCLINE

HYPOLIMNION
cooler

temperature gradient

(b) Autumn

even temperature leads
to vertical mixing

(c) Winter

AIR BELOW 0 °C
ICE

WATER BETWEEN
0 °C AND 4 °C

ABOUT 4°C

temperature gradient

(d) Spring

even temperature leads
to vertical mixing

Figure 3.8   The thermal stratification and mixing of lakes at different times of the year.

small temperature increase. If external heat sources, particularly solar energy, are providing too much energy to be accommodated by an animal or plant, the vaporization of a small amount of water, called 'transpiration' in plants and 'perspiration' in animals, provides an effective cooling system. The vaporization of water absorbs over 500 times as much heat as does raising the temperature of the same mass of water by 1 K. Vaporization can occur at any temperature, but at the boiling point the tendency to vaporize becomes so great that all the extra heat energy added to the liquid is used to convert it into a gas and the temperature remains constant. The boiling point is a function of the total gas pressure:the higher the pressure, the higher the boiling point. We have already mentioned the effect of evaporation and condensation in reducing climatic differences. The **thermal capacity** (the amount of heat required to raise the temperature by a given amount) of large masses of water tends to stabilize temperatures of nearby land masses by absorbing heat in the summer and releasing it in the winter, so reducing seasonal temperature differences.

A simple liquid **solution** contains two substances mixed together homogeneously in the liquid. One substance, whose physical state as a liquid is preserved in the solution, is called the **solvent**; the other substance, which originally may have been a solid or a liquid or a gas, is called the **solute**. The solution consists of the solute particles dispersed amongst, and surrounded by, the solvent particles (Figure 3.9). The formation of the solution involves breaking the attractive forces between solute particles so that the solute may be dispersed in the solvent. For this dispersion to be stable there must be stronger attractive forces between the solvent and the solute than between solute particles alone or solvent particles alone. When attached to solvent particles, the solute particles are said to be **solvated**. If both substances present are liquids, e.g. water and alcohol, the substance present in

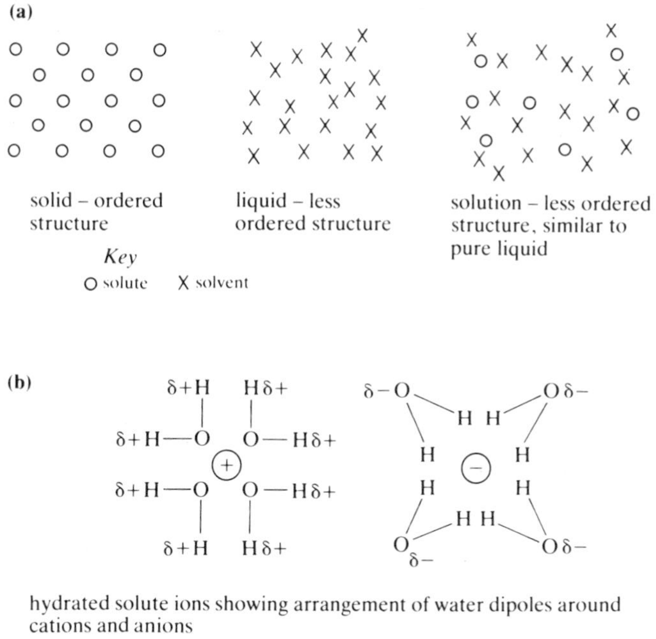

(a)

solid – ordered structure

liquid – less ordered structure

solution – less ordered structure, similar to pure liquid

Key

O solute   X solvent

(b)

hydrated solute ions showing arrangement of water dipoles around cations and anions

Figure 3.9   The relationship between solvent and solute in solutions.

excess is usually called the solvent, but the distinction may not be clear. More complex solutions can be formed by dissolving more than one substance in the solvent.

The solvent properties of water are determined by its dipolar nature and by the presence of hydrogen bonds. The solute particles in an aqueous solution are said to be **hydrated**. In the case of ionic substances the dipolar nature of water makes it particularly suitable as a solvent because the negative end of the dipole, oxygen, is attracted to the positive ions, and the positive ends of the dipole, hydrogen, are attracted to the negative ions (Figure 3.10). The ions at the edges of ionic solids are less strongly held in the crystal structure as they are not completely surrounded by other ions: their electro-static fields will attract water molecules. If enough water molecules surround the ion they will pull it away from the surface of the solid,

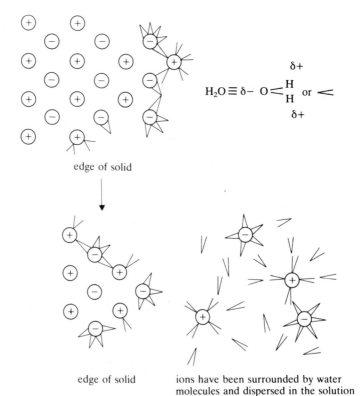

Figure 3.10   The mechanism by which ionic solids dissolve in water.

because their combined attractive forces will be greater than the forces holding the ion in the crystal. The solute ion will become completely hydrated, moving freely through the solvent and away from the dissolving solid. The removal of some ions leaves other ions exposed and these are then hydrated and move away. Eventually either all the solid dissolves or there are so many hydrated ions in the solution that no more can be accepted and the solution is said to be **saturated**. In a saturated solution as many hydrated ions leave the solid as are returned and a state of **dynamic equilibrium** exists. Even when a solid is dissolving some of the hydrated ions are being redeposited on the surface, but the rate of removal is more rapid than the rate of return so that eventually it may all dissolve.

Polar covalent molecules, e.g. ethanol (C_2H_5OH), which are similar to water in that they have a small positive and a small negative charge due to electronegativity differences between the oxygen and hydrogen of the OH group, dissolve in water by a similar mechanism. However, the strength of the solvent–solute interaction is less than with ionic compounds. Non-polar covalent compounds, e.g. methane (CH_4), do not attract water molecules very strongly and usually the water–water interaction is stronger than the water–solute interaction, so the non-polar covalent compound does not dissolve.

When writing the formulae of compounds that are in aqueous solution we do not normally specify the fact that each solute species is hydrated. If we want to emphasize this hydration, the symbol '(aq)' is often used – the formula $Na^+_{(aq)}$ $Cl^-_{(aq)}$, for example, indicates that the sodium chloride is dissolved in water and exists as a collection of hydrated sodium ions and hydrated chloride ions. The number of water molecules directly attached to a given species is determined by the relative size of the water molecules and the species, its charge and the filling of its electron energy levels. In most cases there are either four or six water molecules in the hydrated species.

The solubility of most solids increases with a rise in temperature of the aqueous solution, but the solubility of most gases decreases with a rise in temperature. This latter effect is of great importance with respect to the amount of dioxygen dissolved in water. One of the problems with using river water as a coolant for industrial processes or in electricity power stations is that the hotter water returned to the river contains less dissolved oxygen and so is less able to support aquatic life or to oxidize dead organic matter.

Liquid water exists as an equilibrium mixture of undissociated water molecules, H_2O, hydrogen ions, H^+, and hydroxide ions, OH^- (Eqn 3.15).

$$H_2O \rightleftharpoons H^+ + OH^- \qquad (3.15)$$

The symbol '$\rightleftharpoons$' indicates that this is an **equilibrium reaction** with a constant interchange between the water molecules and the hydrogen and hydroxide ions, but there is a fixed relationship between the quantities of species on the left- and right-hand sides of the equation. For dilute solutions the equilibrium constant is a function of the **molar concentrations** of the species (Eqn 3.16).

$$\text{equilibrium constant, } K = \frac{[H^+][OH^-]}{[H_2O]} = 1.8 \times 10^{-16} \text{ mol dm}^{-3} \text{ at } 25\,°C \quad (3.16)$$

where [ ] indicates the molar concentration of the species inside the square brackets.

The molar concentration is expressed as the **molarity**, symbol M, and is the number of moles of the species dissolved in 1 dm³ (1 litre) of solution. One mole contains **Avogadro's number** of units of the species. If the species are atoms, 1 mole contains the relative atomic mass in grams, e.g. 4 g of $_2^4He$; if the species are molecules, 1 mole contains the relative molecular mass in grams, e.g. 2 g of H_2; if the species are ions, 1 mole contains the relative ionic mass in grams, e.g. 17 g of OH^-. In many cases, especially in more concentrated solutions, there are interactions between the solute molecules so that they are not independent of each other in the solution. Rather than using the absolute molar concentrations, the properties of the solution are better described using the activity of the solutes, i.e. the effective concentrations of the solutes taking the interactions into account. The equilibrium constant is expressed in the same way as in Equation 3.16, but the square brackets, [ ], indicate activities rather than molarities.

Undissociated water must be present in great excess because the value of the equilibrium constant, 1.8×10^{-16} mol dm⁻³, is so very small. This means that any changes in the concentration of the water molecules will be negligible and the concentration will be essentially constant at

$$55.56 \text{ mol dm}^{-3} \left(\frac{\text{weight of 1 dm}^{-3} H_2O}{\text{relative molecular mass, } H_2O} = \frac{1\,000}{18} \right)$$

Using this value, Equation 3.16 becomes:

$$K = \frac{[H^+][OH^-]}{55.56 \text{ mol dm}^{-3}} = 1.8 \times 10^{-16} \text{ mol dm}^{-3} \quad (3.17)$$

$$K_w = [H^+][OH^-] = 55.56 \times 1.8 \times 10^{-16} \text{mol}^2 \text{ dm}^{-6}$$

$$= 1 \times 10^{-14} \text{ mol}^2 \text{ dm}^{-6} \text{ at } 25\,°C \quad (3.18)$$

The constant, K_w, is called the ionic product of water. In pure water the concentrations of H^+ and OH^- must be equal, as they can only be formed by the dissociation of the water (Eqn 3.15). When the concentration of the two ions is equal, the solution is said to be neutral, and from Equation 3.18 we see that the concentration of each is 1×10^{-7} mol dm^{-3} (Eqn 3.19).

$$[H^+] = [OH^-] = 1 \times 10^{-7} \text{ mol dm}^{-3} \text{ at } 25\,°C \tag{3.19}$$

To avoid having to keep writing exponential numbers, the hydrogen-ion concentration, and hence the hydroxide-ion concentration, is conveniently expressed using the pH scale (Eqn 3.20).

$$pH = -\log[H^+] \tag{3.20}$$

A neutral solution therefore has a pH of 7. If we also define $pOH = -\log[OH^-]$ and $pK_w = -\log K_w$ and substitute these forms in Equation 3.18, we obtain

$$pK_w = pH + pOH = 14 \tag{3.21}$$

The sum of pH and pOH must always be 14 (at $25\,°C$) in any aqueous solution. In a neutral solution $pOH = pH = 7$ and in general $pOH = 14 - pH$ and $pH = 14 - pOH$. If the hydrogen-ion concentration is greater than the hydroxide concentration, i.e. if the pH is less than 7, the solution is said to be **acidic**. If the hydroxide-ion concentration is greater than the hydrogen-ion concentration, pH greater than 7, the solution is said to be **basic**, or **alkaline**. A broader definition of acids and bases that is not restricted to aqueous solutions is that an acid is a proton donor and a base is a proton acceptor. A proton is a hydrogen ion that has not been hydrated.

Returning to aqueous solutions, because almost all acid–base reactions of environmental interest occur in the presence of water, we find that the concentration of hydrogen ions does not always equal the concentration of acid dissolved. In the case of hydrochloric acid, HCl, which is present in the human stomach at a concentration of about 0.1 M (1×10^{-1} mol dm^{-3}), the pH of such a molarity is 1, i.e. the hydrochloric acid is completely dissociated.

$$HCl \longrightarrow H^+ + Cl^- \tag{3.22}$$

However, ethanoic acid (acetic acid, CH_3COOH), found in vinegar, is only partially dissociated in water (Eqn 3.23) and only about 1% of the molecules have liberated hydrogen ions.

$$\underset{\substack{\text{ethanoic} \\ \text{acid}}}{CH_3COOH} \rightleftharpoons \underset{\substack{\text{ethanoate ions} \\ \text{(acetate ions)}}}{H^+ + CH_3COO^-} \tag{3.23}$$

If the concentration of ethanoic acid in water is 0.1 M, the concentration of H^+ will be about 1×10^{-3} M ($0.1 \times 1 \times 10^{-2}$ M) and the pH will be 3. Acids like ethanoic acid that do not ionize completely in water are called 'weak' acids; acids like hydrochloric acid, that do ionize completely, are called 'strong' acids. Bases can also be divided into 'strong' bases that ionize completely, e.g. sodium hydroxide (NaOH, Eqn 3.24), and 'weak' bases that do not ionize completely, e.g. ammonium hydroxide (NH_4OH, Eqn 3.25).

$$NaOH \longrightarrow Na^+ + OH^- \qquad \text{pH of 0.1 M soln} = 13 \qquad (3.24)$$

$$NH_4OH \longrightarrow NH_4^+ + OH^- \qquad \text{pH of 0.1 M soln} = 11.1 \qquad (3.25)$$

When an acid reacts with a base the products are water and a salt (Eqn 3.26), a compound containing the cation from the base and the anion from the acid.

$$\underset{\substack{\text{sulphuric}\\\text{acid}}}{H_2SO_4} + \underset{\substack{\text{sodium}\\\text{hydroxide}}}{2NaOH} \longrightarrow 2H_2O + \underset{\substack{\text{sodium}\\\text{sulphate (a salt)}}}{Na_2SO_4} \longrightarrow 2Na^+ + SO_4^{2-}$$

$$(3.26)$$

With strong acids and bases the resulting solution will be neutral, pH 7, when equivalent amounts of the two react and when there is no excess of the acid or base. When weak acids react with strong bases and when strong acids react with weak bases, the equilibrium point does not occur at pH 7. The salts that are formed react with water to liberate hydroxide ions, if they are salts of a weak acid and strong base (Eqn 3.27), or hydrogen ions, if they are salts of a strong acid and weak base (Eqn 3.28).

$$\underset{\substack{\text{sodium}\\\text{ethanoate}}}{NaOOCCH_3} + H_2O \rightleftharpoons Na^+ + CH_3COO^-$$

$$+ H_2O \rightleftharpoons Na^+ + \underset{\substack{\text{undissociated}\\\text{acid}}}{CH_3COOH} + OH^- \qquad (3.27)$$

$$\underset{\substack{\text{ammonium}\\\text{chloride}}}{NH_4Cl} + H_2O \rightleftharpoons Cl^- + NH_4^+$$

$$+ H_2O \rightleftharpoons Cl^- + \underset{\substack{\text{undissociated}\\\text{base}}}{NH_4OH} + H^+ \qquad (3.28)$$

These reaction types (Eqns 3.27 and 3.28) are often called hydrolysis because they involve water molecules. Many of the compounds found in natural systems are made from weak acids or bases.

A number of compounds are insoluble in water but will dissolve in either an acid or a base. A compound that dissolves in an acid is said to be basic and a compound that dissolves in a base is said to be acidic. Most metals have basic oxides, forming alkaline solutions or dissolving in acids, and most non-metals have acidic oxides, forming acid solutions or dissolving in bases.

Because the majority of naturally occurring acids and bases are weak, being only partially ionized in aqueous solutions, the range of pH is relatively small. The range is also restricted by the action of **buffers**, which are weak acids or weak bases in the presence of their salts. The buffer solution maintains a reasonably constant pH, even when relatively large amounts of either acid or base are added. The carbon dioxide–bicarbonate–carbonate system is a particularly important buffer system regulating the pH of the oceans and of blood. The mechanism by which this and other buffer systems operate will be discussed in Chapter 4.

One reason why oxides of sulphur and nitrogen have such a great effect on the pH of rain is that both sulphuric acid, H_2SO_4, and nitric acid, HNO_3, are strong acids. Thus the concentration of hydrogen ions liberated is maximized by the complete ionization of the acids.

4 Carbon

Carbon Abundance by weight (the relative abundance is given in parentheses): Earth 350 ppm (14); crust, 200 ppm (17); ocean, inorganic, 28 ppm (10), organic, 2 ppm (15); atmosphere as CO_2, 354 ppm (4); as CH_4, 1.5 ppm (7). ppm = $mg\,kg^{-1}$.

4.1 The carbon cycle

Living organisms are mainly composed of water and various carbon compounds; therefore the cycling of carbon is of prime importance to the support of life. We have already seen that the cycling of one element does not occur in isolation from other elements and some aspects of importance to the cycling of carbon are dealt with in other chapters.

The carbon cycle is outlined in Figure 4.1. The majority of the carbon is found in rocks, either as carbonate (CO_3^{2-}), usually associated with calcium in limestones (as $CaCO_3$), or as dispersed organic carbon in sedimentary rocks, particularly shales. The carbonates are described as containing inorganic carbon, with about three-quarters of the total carbon in the outer regions of the Earth in these inorganic deposits and one-quarter in dispersed organic compounds. The combined carbon content of all the other reservoirs (atmosphere, land **biota**, soil humus, fossil fuels, marine biota, dissolved compounds) comes to less than 1% of the total.

4.2 Aqueous systems

The ocean is a much more complex system than is indicated by the single reservoir in Figure 4.1. Water is a good absorber of light and there is only sufficient radiation to support photosynthesis down to about 200 m below the surface. However, light is not generally the limiting factor for primary production in the surface layers of the sea. The concentrations of nitrogen, phosphorus, silicon and some essential trace elements are the major factors determining marine productivity. The total marine biomass is much smaller (200 times less) than

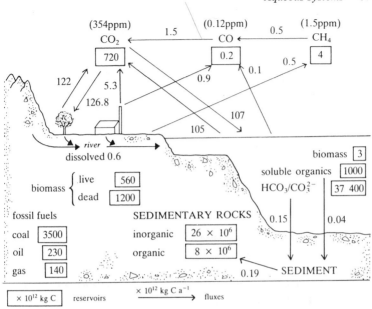

Figure 4.1 The carbon cycle.

the terrestrial biomass, reflecting the low productivity of the majority of surface ocean layers and the great variability from one area to another. The most fertile areas are of very limited extent, being confined to narrow bands around a number of continental coasts and the seas near Antarctica. In each of these regions there are upwelling currents, bringing higher concentrations of nutrients to the surface. The majority of the world's oceans have productivities which are four to ten times lower than the most fertile areas. The majority of the oceans' volume is unproductive because of the lack of light. In addition, there is relatively little vertical mixing and this prevents the influx of new nutrients into the productive surface layers.

The primary **producers** in the oceans are unicellular algae called phytoplankton, which are capable of photosynthesis. These and the zooplankton that live on them (Figure 4.2) are capable of only limited locomotion, and drift with the ocean currents. The zooplankton and phytoplankton provide food for the organisms called **nektons**, such as fish and whales, that are capable of locomotion and for the bottom-dwelling organisms, called **benthos**. The average residence time for the carbon in oceanic biota is less than a month, indicating the rapid turnover that occurs. The gradual settling out of some of the dead

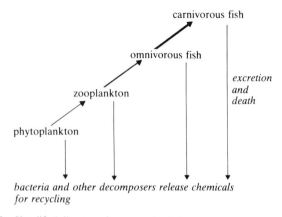

carnivorous fish

omnivorous fish

*excretion
and
death*

zooplankton

phytoplankton

*bacteria and other decomposers release chemicals
for recycling*

Figure 4.2   Simplified diagram of an ocean food chain.

remains of these various organisms leads to the incorporation of organic compounds in sediments.

The major carbon reservoir in the oceans is the so-called inorganic carbon. Carbon dioxide dissolves in water to give a hydrated molecule, $CO_{2(aq)}$, which then forms an equilibrium mixture containing bicarbonate (hydrogencarbonate, HCO_3^-) and carbonate (CO_3^{2-}) ions. At pHs lower than those found in sea water, carbonic acid (H_2CO_3) will also be present.

$$CO_{2(gas)} \overset{water}{\rightleftharpoons} CO_{2(aq)} \tag{4.1}$$

$$H_2O + CO_{2(aq)} \rightleftharpoons H_2CO_3 \tag{4.2}$$

$$H_2CO_3 \rightleftharpoons H^+ + HCO_3^- \tag{4.3}$$

$$HCO_3^- \rightleftharpoons H^+ + CO_3^{2-} \tag{4.4}$$

Most oceanic waters have a pH in the range 8 to 8.3, as they contain more hydroxide ions than hydrogen ions due to reactions 4.5 and 4.6, and the carbonate–bicarbonate mixture contains about 13% carbonate.

$$HCO_3^- \rightleftharpoons CO_{2(aq)} + OH^- \tag{4.5}$$

$$CO_3^{2-} + H_2O \rightleftharpoons HCO_3^- + OH^- \tag{4.6}$$

When carbon dioxide dissolves in sea water the overall reaction can be summarized by Equation 4.7.

$$CO_{2(aq)} + H_2O + CO_3^{2-} \rightleftharpoons 2HCO_3^- \tag{4.7}$$

The concentration of the various components varies with depth. In the surface layers where photosynthesis is active and CO_2 is being consumed, the reaction (Eqn 4.7) moves to the left. Equilibrium reactions have a constant relationship between the concentration of reactants on the left-hand side of the equation and the concentration of the products on the right-hand side of the equation. The equilibrium constant, K_{eqm}, for reaction 4.7 is

$$K_{eqm} = \frac{[HCO_3^-]^2}{[CO_{2(aq)}][H_2O][CO_3^{2-}]} \qquad (4.8)$$

$$K_{eqm}[CO_{2(aq)}][H_2O][CO_3^{2-}] = [HCO_3^-]^2 \qquad (4.9)$$

If the concentration of dissolved carbon dioxide, $[CO_{2(aq)}]$, is reduced, the concentration of bicarbonate ion, $[HCO_3^-]$, must also be reduced. Removing carbon dioxide causes some of the bicarbonate to be converted to CO_2, H_2O and CO_3^{2-}, until the concentration products are again the same on both sides of Equation 4.9, i.e. reaction 4.7 moves to the left.

In deeper water there is a net production of carbon dioxide due to respiration and decay processes which oxidize the organic compounds, so outweighing photosynthesis. Under these conditions reaction 4.7 moves to the right as the extra carbon dioxide is partially converted to bicarbonate and equilibrium is again attained.

These changes in concentration of CO_2, HCO_3^- and CO_3^{2-} affect the solubility of calcium carbonate, $CaCO_3$. The surface waters contain higher concentrations of carbonate as reaction 4.7 moves to the left and the solubility product, K_{sp}, of calcium carbonate is exceeded (Eqns 4.10 and 4.11), and any calcium carbonate present will not dissolve.

$$CaCO_{3(solid)} \rightleftharpoons Ca^{2+}_{(aq)} + CO^{2-}_{3(aq)} \qquad (4.10)$$

$$K_{sp} = [Ca^{2+}][CO_3^{2-}] = 4.47 \times 10^{-8}\,mol\,dm^{-3} \qquad (4.11)$$

As the depth increases, the concentration of carbonate ion decreases and calcium carbonate becomes soluble. Shells will therefore start to dissolve.

The presence of the various inorganic carbon species plays a particularly important part in controlling the pH of natural waters.

Buffer solutions and alkalinity

The ocean is described as being 'buffered' because relatively large quantities of acid or base can be added to sea water without causing much change in the pH. Many freshwater lakes and rivers are not buffered and their pHs change rapidly with the addition of acid or base. Even the sea is not able to act as a buffer indefinitely and eventually there will be a marked change in pH when the buffering capacity is exhausted. The buffer system in the ocean is quite complex.

Added acids cause the equilibrium points of reactions 4.1–4.4 to move to the left, and added bases move the equilibria in reactions 4.5 and 4.6 to the left; in each case the number of free hydrogen or hydroxide ions is reduced. Sea water also contains phosphates, silicates and other species, especially borates (Eqn 4.12), which act as buffers.

$$H_3BO_3 \rightleftharpoons H^+ + H_2BO_3^- \rightleftharpoons H^+$$
$$+ HBO_3^{2-} \rightleftharpoons H^+ + BO_3^{3-} \quad (4.12)$$

Because of the large number of equilibrium reactions to be considered, calculating the buffering capacity of sea water is very difficult. This difficulty also applies to other natural waters. A practical solution to the problem has been found. The capacity of the natural water to neutralize acid is called the alkalinity. The alkalinity is determined by titration of the water with a strong acid (Figure 4.3). By actually determining the alkalinity curve using a pH meter and the addition of the strong acid, the effects of all the species in the water can be taken into account. The carbonate buffer system is the major factor controlling the ability of most natural water systems to withstand a large change in pH when acid is added. The longer the section CD (Figure 4.3), the more stable the water will be. The two sharp increases in slope correspond to the reactions 4.4 (BC, Figure 4.3) and 4.3 (DE, Figure 4.3). For most natural water systems these changes occur at pH 8–8.5 and pH 4–5, respectively.

Fish and many other organisms are unable to survive large drops in pH. The salmonids (e.g. salmon and trout) are particularly susceptible, and fish kills can occur at pHs below 4.5–5. The greater the alkalinity of the water, i.e. the larger the amount of acid required to reach section DE (Figure 4.3), the less likely are harmful effects due to pH changes. One of the reasons for the variability in the effects of acid rain is due to the varying alkalinity of lakes and rivers. Scandinavian freshwater systems tend to have low alkalinities because the underlying rocks do not provide many carbonate or other buffering species. Therefore, small additional quantities of acid cause large pH changes. Similarly, industrial acidic effluents pumped into a high-alkalinity river may cause little upset whereas in a low-alkalinity river the same effluent could be devastating.

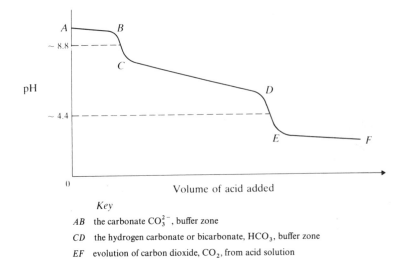

Key

AB   the carbonate CO_3^{2-}, buffer zone

CD   the hydrogen carbonate or bicarbonate, HCO_3, buffer zone

EF   evolution of carbon dioxide, CO_2, from acid solution

Figure 4.3   An idealized diagram of the alkalinity curve obtained when changes in pH are plotted against the volume of acid added to an aqueous carbonate solution.

4.3 Photosynthesis and the formation of carbon compounds

We have already briefly discussed photosynthesis as a source of dioxygen and as a means of storing solar energy as chemical energy (Chapter 2). The detailed mechanism of the process is very complex and has not yet been fully unravelled. However, many of the changes taking place are now known.

The light is absorbed by a number of pigment molecules, the most abundant of which are chlorophyll a and b. In the higher plants the pigments are concentrated in the green-tissue cells. Chlorophyll molecules contain a magnesium ion surrounded by a porphyrin-ring structure (Figure 4.4). The porphyrin-ring structure is found in a number of biologically important molecules, such as haemoglobin, vitamin B_{12} and the cytochromes. All the porphyrins have four nitrogen-containing pyrrole rings but with a variety of side chains. The difference between chlorophyll a and chlorophyll b is due to different groups in position X (Figure 4.4). The presence of the alternating single and double bonds, called a conjugated double-bond system, allows the electrons in these bonds to be relatively easily raised in energy. There are thought to be two pigment systems in the green cells and each is responsible for absorbing particular parts of the light spectrum.

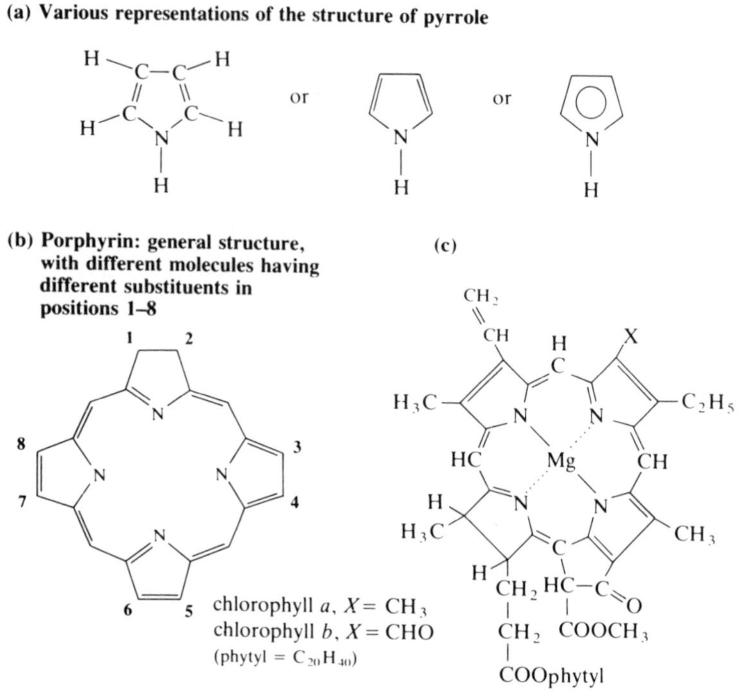

(a) Various representations of the structure of pyrrole

(b) Porphyrin: general structure, with different molecules having different substituents in positions 1–8

(c)

chlorophyll *a*, X = CH$_3$
chlorophyll *b*, X = CHO
(phytyl = C$_{20}$H$_{40}$)

Figure 4.4   Structures of (a) pyrrole, (b) a porphyrin ring, (c) chlorophyll.

One system, containing chlorophyll a, absorbs red light ($\lambda \simeq 650$ nm, energy $\simeq 185$ kJ mol^{-1}) to produce excited molecules containing electrons in higher energy states. These electrons are transferred to compounds such as NADPH or ATP.

ATP is the contraction for adenosine-5′-triphosphate, a molecule used in biological systems to provide the energy required to drive reactions needing an energy input. It is a very important molecule, being the major energy carrier in metabolic processes. The ATP transfers a phosphate group, PO$_4^{3-}$, to one of the organic reactants (Eqn 4.13), which is then able to react in the desired manner (Eqn 4.14). The addition of phosphate is called a **phosphorylation** reaction.

$$\underset{\text{organic molecule}}{\text{ROH}} + \text{ATP} \longrightarrow \underset{\substack{\text{phosphorylated} \\ \text{intermediate}}}{\text{ROPO}_3} + \underset{\text{adenosine-5′-diphosphate}}{\text{ADP}} \qquad (4.13)$$

$$\text{ROPO}_3 \xrightarrow[\text{reaction}]{\text{further}} \underset{\text{organic product}}{\text{ROX}} + \underset{\text{phosphate}}{\text{PO}_4^{3-}} \qquad (4.14)$$

ATP can be regenerated from ADP by the addition of a phosphate group. As the system is more or less cyclic (Figure 4.5), ATP acts as an efficient energy-transfer agent. The energy may come directly from photosynthesis or from the chemical energy in compounds such as carbohydrates and fats that are broken down as required to regenerate ATP.

NADPH is reduced nicotinamide adenine dinucleotide phosphate and acts as a carrier of protons and electrons (Eqn 4.15).

$$\text{NADPH} + \text{oxidized compound} \longrightarrow \text{NADP}^+ + \text{reduced products} \quad (4.15)$$

The NADPH can be regenerated from NADP^+, nicotinamide adenine dinucleotide phosphate (Eqn 4.16).

$$\text{NADP}^+ + \text{H}^+ + 2e^- \longrightarrow \text{NADPH} \quad (4.16)$$

During photosynthesis the protons and electrons are provided by water with the release of dioxygen (Eqn 4.17).

$$2\text{H}_2\text{O} \longrightarrow \text{O}_2 + 4\text{H}^+ + 4e^- \quad (4.17)$$

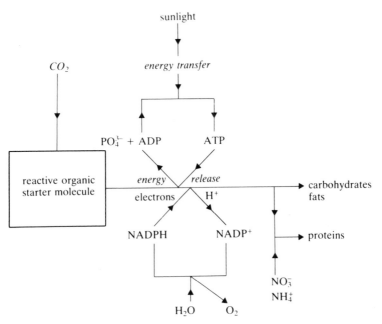

Figure 4.5 Diagrammatic representation of the use of ATP and NADPH in a photosynthetic reaction sequence.

The incorporation and reduction of carbon dioxide to form the various organic products of photosynthesis involves a complex series of cyclic reactions (Figure 4.5). One molecule of carbon dioxide is added to an activated organic molecule. The product is then progressively reduced by NADPH, and rearranged, under the influence of enzymes, until the final compound, whether a carbohydrate, lipid or protein, is produced. When an energetically unfavourable reaction has to occur, the driving force is provided by ATP injecting energy by means of a phosphorylation reaction. The enzymes provide conditions that favour very specific changes, whether additions, rearrangements or decompositions. They increase the rate of the reaction, but are not consumed in the reaction, and are therefore catalysts.

The consequences of photosynthesis are that the **photoautotrophs** are net *absorbers* of carbon dioxide, water and sunlight, and net *producers* of organic molecules and dioxygen. The photoautotrophs use some of the organic compounds as sources of energy for many of the reactions carried out in their cells. Energy stored as chemical compounds can be transported and released as required, so increasing the efficiency of the organism because sunlight is not available during the night, all cells need not be exposed to sunlight, and there may be temporary shortfalls in the availability of carbon dioxide and water. The energy-releasing process (Eqn 4.18) is called aerobic respiration and may be thought of as the reverse of photosynthesis.

$$(CH_2O)n + nO_2 \longrightarrow nCO_2 + nH_2O + \text{energy} \qquad (4.18)$$

The respiration process, like photosynthesis, occurs by a complex series of reactions involving many steps and the use of enzymes. The energy released by a reaction like the oxidation of glucose (Eqn 4.19) is not given out all at once, but rather as smaller packets mainly to regenerate ATP and thus drive the desired reaction steps in the life processes of the organism.

$$C_6H_{12}O_6 + 6O_2 \longrightarrow 6H_2O + 6CO_2 + \text{energy} \qquad (4.19)$$

Heat is also produced and this may help the metabolic processes, as at higher temperatures the rate of chemical reactions can be increased quite significantly. For many reactions involving organic compounds a 10 K increase in temperature will double the rate of reaction. These higher rates of reaction will allow more energy to be produced in a shorter time thus enabling many more energy-consuming reactions to take place without running short of energy. By proper use of the heat-transfer properties of aqueous solutions, the organism can control the environment of its cells and maintain them at a higher temperature than the external surroundings. This control produces a more stable internal environment that can be optimized to suit the needs of the organism. The life-forms that have taken greatest advantage of the flexibility allowed by utilizing chemical compounds and respiration as energy sources are the **heterotrophs**. The photoautotrophs convert the solar energy into 'high energy' chemical compounds. The excess compounds which they do not use for their own metabolic processes can be consumed by heterotrophs. The result has been the development of **food chains** such as in Figure 4.2, in which the primary producers utilize solar energy directly and the various **consumers** depend upon either the producers or other consumers for their energy intake. The actual carrying out of the photosynthesis process uses up a lot of the energy extracted from the Sun by the plants, etc. The consumers do not carry this energy burden and can use the energy in the already synthesized molecules for their own growth, biosynthetic needs and the other metabolic processes that control their internal environment and permit muscular activity.

Carbohydrates and lipids

The two major groups of energy-storing chemicals produced by photosynthesis are carbohydrates and lipids. One important source of biologically available energy is provided by the breakdown of monosaccharides such as glucose (Eqn 4.19). In the presence of dioxygen, the complete decomposition of the glucose to carbon dioxide and water can be achieved with the release of the maximum amount of energy ($2880 \, kJ \, mol^{-1}$) accumulated during photosynthesis. The majority of this energy is transferred inside the organism by the 38 molecules of ATP that are synthesized during this aerobic respiration. The first steps in the oxidation of glucose lead to the formation of pyruvic acid ($CH_3COCOOH$) and enough energy to generate two ATP molecules. Up to this stage the presence of dioxygen is not required and the reaction steps are common to both aerobic respiration (oxygen dependent) and **anaerobic** respiration or fermentation (oxygen independent). Aerobic respiration leads to carbon dioxide and water production, but with anaerobic respiration there is no suitable electron acceptor and all the carbon atoms in the pyruvic acid cannot be converted into carbon dioxide. Organisms such as yeast produce ethanol (ethyl alcohol, C_2H_5OH) and carbon dioxide in the best known of the fermentation processes. The souring of milk is another fermentation process in which lactic acid ($CH_3CH(OH)COOH$) is formed. Those organisms that obtain their energy by the fermentation pathway are at a great disadvantage compared to those using aerobic respiration because of the relatively small amount of energy released during fermentation. With glucose, fermentation releases $220\,kJ\,mol^{-1}$, whereas aerobic respiration releases $2880\,kJ\,mol^{-1}$.

Glycogen is the polysaccharide that is used as a nutritional reserve by animals. The equivalent compound in plants is starch, which is stored either in the roots of tubers, like potatoes, or in seeds, as with wheat, rice and maize. The other major polysaccharide is cellulose, which contains 30% to 90% of all the carbon in vegetation and provides the main supporting structure as wood or fibre. Whereas glycogen and starch can be used as food by humans, cellulose cannot. The polysaccharides must first be broken down into monosaccharides in order that their energy may be released. Humans, and many other animals, do not have the necessary cellulase enzymes, which break up cellulose, in their digestive tracts. Cellulose consists of glucose units joined together by β-linkages (Figure 4.6), whereas glycogen and starch have their glucose units joined together by α-linkages. This very small change is enough to give the radical difference in digestibility and indicates the importance of **stereochemical** factors in determining reaction pathways.

The four covalent bonds around a carbon atom are directed tetrahedrally in space (Figure 4.7) so that they are as far from each other as possible and all of the bond angles are 109.5°. One result of

(a) An alpha, α, linkage – oxygen A 'opposite' oxygen B

(b) A beta, β, linkage – oxygen A 'adjacent to' oxygen B

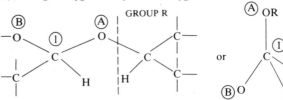

Figure 4.6 Alpha and beta linkages in polysaccharides.

these directional characteristics is that when four different groups are attached to a carbon atom there are two possible compounds that have identical constitutional formulae, but different spatial arrangements. These two spatial arrangements, called configurations, are mirror images of each other and they are an example of **stereoisomerism**. As can be seen from Figure 4.7, but probably better with three-dimensional models, the two configurations are not superimposable, no matter how they are rotated. The type of stereoisomerism shown by these molecules with no plane of symmetry is called optical isomerism, because the two isomers rotate **plane polarized light** in opposite directions. Many organic compounds, such as carbohydrates, have at least one asymmetric carbon atom, with four different groups attached to it. The different configurations can influence whether the stereoisomers react, particularly if enzymes are required to promote the reaction.

Enzymes are proteins that have the ability to speed up organic reactions without being 'used up' themselves. As well as the proteins, the enzyme molecule usually contains other chemical species, including metals. Both the protein and the other species must be present for the enzyme to be active; neither alone is sufficient. Enzymes have different degrees of specificity. Many will only catalyse one particular reaction of one particular compound. Others will catalyse reactions of a certain bond type, e.g. a double bond, no matter what other groups are present. The remaining enzymes lie between these two extremes.

The enzymes have 'active sites' at which the reaction occurs. The remainder of the enzyme molecule helps to keep the molecule stable

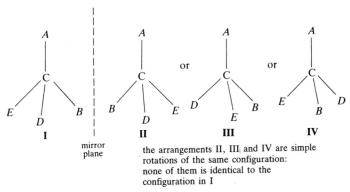

Figure 4.7 Stereoisomerism exhibited by a carbon atom tetrahedrally attached to four different groups.

and to provide the correct configuration (spatial arrangement of groups) at the active site. The enzyme will increase the rate at which the reaction will reach equilibrium, but it will not change the equilibrium position; it speeds up the reaction, but does not increase the theoretical yield. The molecule that is to undergo reaction is called the substrate: this becomes attached to the enzyme at the active site (Figure 4.8). The relative shapes of the substrate and the active site decide the degree of specificity of the enzyme. In some cases the stereochemical fit has to be much closer than in others. The bound substrate is now held in such a way that the atoms that are to take part in the reaction are in a favoured position for interaction with the incoming reacting species (Figure 4.8b).

A reaction is favoured if the total energy in the chemical products is lower than the total energy in the reactants (Figure 4.9). For many

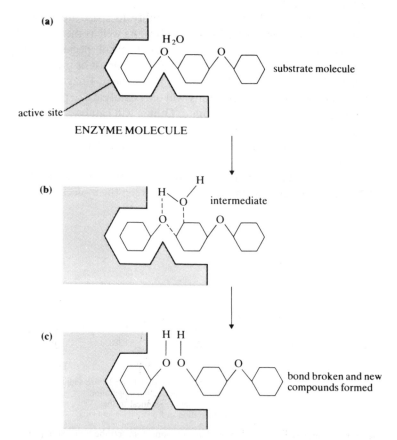

Figure 4.8   A representation of how an enzyme can influence a reaction mechanism.

(a) **The initial, *I*, and final, *F*, energy states for an energetically favourable reaction**

(b) **Because chemical bonds may need to be broken before the reaction can proceed, an intermediate excited (higher-energy) state, H_1, occurs – the energy required to reach this intermediate state is called the activation energy, A_1**

(c) **The presence of an enzyme can introduce the possibility of a number of enzyme-plus-substrate intermediate stages (E_1, E_2), which lead to a series of small activation energies (A_2, A_3, A_4), each less than A_1**

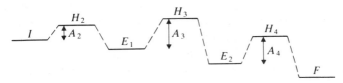

Figure 4.9 The effect of enzymes on the activation energy of a reaction.

reactions, bonds have to be broken before reactions can proceed; the initial overcoming of these bonding forces requires an input of energy called the **activation energy**. The activation energy acts as a barrier to the completion of the reaction: there must be this energy input before there is the energy output from the formation of the final product (Figure 4.9b). The larger the activation energy the more difficult it will be for the reaction to proceed. Enzymes provide an alternative pathway comprising several steps, each of which has a relatively small activation energy; molecules following this pathway can react more rapidly. In many cases there is no suitable enzyme and the activation energy is too great for the reaction to proceed. The breakdown of the polysaccharides in animals' digestive tracts depends upon the presence of the enzymes required to reduce the effects of the activation-energy barriers. If the correct enzymes (e.g. cellulase for cellulose) are not present, no breakdown occurs.

Plants use starch as their main energy-storage compound, but animals use fats. The triglyceride fats contain more material capable of being oxidized than do carbohydrates of the same weight. The maximum energy that an organism can obtain from a fat is about twice as much as from carbohydrate, on a weight-for-weight basis. Fats, therefore, are more concentrated energy-storage compounds than carbohydrates. Carbohydrates can be converted to fats, and when the intake of food is greater than immediate needs, large quantities of fat can be deposited around the body. Other properties of fats that make them particularly suitable as storage compounds are that they are insoluble in water and that the presence of large quantities does not upset the ionic balance in the aqueous solutions of the body.

4.4 Energy

The use of various fuels and their conversion into different energy forms (thermal, electrical, chemical, mechanical, etc.) has to be considered with respect to the 'usefulness' of these energy forms and the laws of thermodynamics that control the operation of energy systems. The laws of thermodynamics only deal exactly with systems in equilibrium. If applied to non-equilibrium systems, approximations must be made, but these are often relatively unimportant. Statements of the various laws and paraphrases of these are given in Table 4.1.

Energy is generally defined as the 'capacity to do work', and the laws of thermodynamics indicate how much useful energy (work) we can obtain from a system. With every energy conversion (e.g. chemical to thermal, mechanical to electrical) there is a net loss of energy from

Table 4.1   The laws of thermodynamics paraphrased (after Walters, G. A. and E. M. Wewerka 1974. *Contemporary chemistry.* Columbus OH: Charles E. Merrill). The so-called paraphrases are expressions of how the operation of the laws of thermodynamics limits the extraction of usable energy from a system

Law	Statement	Paraphrase
zeroth	there exists a phenomenon called temperature	life is a game (there are rules)
first	energy is neither created nor destroyed in a process but merely transformed from one form to another	you cannot win; the best you can do is break even
second	the entropy of a system and its surroundings increases for any spontaneous process	you can only break even if you live for ever
third	absolute zero of temperature is approachable, but unattainable	you cannot live for ever

the system. There can never be a 100%-efficient conversion of energy. This loss in 'capacity to do work' shows up as an increase in **entropy** (symbol S). Entropy can be thought of as the tendency for a system to achieve equilibrium, i.e. its most probable state of energy distribution. This implies that the system will become disordered, because an ordered arrangement has a lower probability than a disordered arrangement provided that there is free movement from one arrangement to another. Consider the example of three coins that have been spun (Figure 4.10a). There is one chance that they will land all heads and one chance that they will be all tails, i.e. two chances of these completely ordered arrangements. There are six chances that some other arrangement will occur. Similarly, the spinning of four coins (Figure 4.10b) could give two arrangements with the highest possible order (all heads, or all tails) and 30 other arrangements showing various degrees of lower order. In general, the total number of arrangements is 2^n, where n is the number of coins (Figure 4.10). However, once formed, the ordered arrangement of all ten coins being heads or tails will remain until they are spun again. The ground or table prevents the coins from continuing to spin and eventually reaching their thermodynamically most probable equilibrium state. Similarly, ordered systems are common throughout the Universe, preserved in their ordered arrangement until the restraining or binding forces that hold them together are broken.

Chemical reactions and phase changes (solid$\rightleftharpoons$liquid$\rightleftharpoons$gas) involve heat changes. If the change occurs under conditions of constant

(a) Three coins

H H H,

H H T, H T H, T H H,

H T T, T H T, T T H

T T T

Total number of arrangements $= 2^3 = 8$, with 2 out of 8 having the highest degree of order (i.e. all heads or all tails)

(b) Four coins

H H H H

H H H T, H H T H, H T H H, T H H H

H H T T, H T H T, T H H T

T T H H, T H T H, H T T H

H T T T, T H T T, T T H T, T T T H

T T T T

Total number of arrangements $= 2^4 = 16$, with 2 out of 16 having the highest degree of order

Figure 4.10 The possible arrangements that may result from spinning (a) three, (b) four coins.

pressure, the heat absorbed or evolved is called the **enthalpy** change, ΔH. The Greek capital letter delta, Δ, is commonly used as the symbol for 'change in value of'. For a reaction in which heat is evolved (given out), an **exothermic** reaction, the value of ΔH is negative: heat is being *lost* from the reaction system. When heat is absorbed (taken in), an **endothermic** reaction, the value of ΔH is positive: heat is being *gained* by the reaction system. Similarly, if the entropy of a system decreases, i.e. if the system becomes more 'ordered' (e.g. if gas condenses to liquid), the value of ΔS is negative. When the entropy of a system increases, i.e. if it becomes more 'disordered', the value of ΔS is positive.

The ability of a reaction to do work is indicated by the **free-energy** change, ΔG, that occurs. The relationship between free-energy change, ΔG, enthalpy change, ΔH, entropy change, ΔS, and absolute temperature, T, is given by Equation 4.20.

$$\Delta G = \Delta H - T \Delta S \qquad \text{at constant temperature and pressure} \qquad (4.20)$$

The more negative the free-energy change, the more energy the system can supply, the more work it can do and the more stable the final products will be. Stability is high if a lot of energy has to be added to the system to reverse the reaction.

Fossil fuels combine with dioxygen, when burnt, to produce products that are thermodynamically more stable than the fuel. The free-energy change for combustion is negative. The chemical energy is released as heat. However, for work to be carried out, the thermal energy must be converted to mechanical energy. The efficiency of this process depends upon the temperature difference between the heat source before the conversion has taken place and the temperature after the energy has been extracted (Eqn 4.21).

$$\text{Carnot efficiency} = \text{efficiency of energy conversion}$$

$$= \frac{T_{(in)} - T_{(out)}}{T_{(in)}} \qquad (4.21)$$

where $T_{(in)}$ = absolute temperature before energy extracted;
$T_{(out)}$ = absolute temperature after energy extracted.

For instance, in an oil-fired power station the hot gases entering the boiler to produce the steam which drives the turbines may heat the steam to 650 K and the water may leave the condenser at 300 K. The efficiency of energy conversion would be about 54%:

$$\left(\frac{650 - 300}{650} \times 100 \right)$$

The efficiency could be increased by raising the input temperature or by lowering the output temperature. The input temperature is limited by (a) the temperature of combustion of the fuel, (b) the design of the furnace/boiler system, and (c) the ability of the construction materials to withstand high temperatures without deterioration. The lower limit is determined by the temperature of the cooling water in the condensers and the ambient air temperature. Modern fossil-fuelled power stations are operating reasonably closely to their Carnot efficiencies for the steam cycle, but much higher overall efficiencies could be obtained if the steam cycle could be taken up to the temperature of the combustion gases, i.e. 1000 °C or 1300 K.

$$\text{Carnot efficiency} = \frac{1300 - 300}{1300} \times 100 = 77\%$$

When geothermal energy is substituted for fossil fuel the input temperatures of the steam from the geothermal field are about 200–250 °C, say 500 K. The Carnot efficiency is then 40%. The lower the difference between possible input and output temperature, the lower the efficiency of the system in terms of the useful work that can be extracted. Any energy source (e.g. fossil fuels and electricity) that can provide high temperatures is considered to be a 'high grade' energy source, whereas sources (e.g. solar water heaters and geothermal energy) that can only produce low temperatures are considered to be 'low grade' energy sources.

Solar radiation consists of two components. One component is the direct radiation received when the Sun is visible. This direct radiation can be focused so that the energy is concentrated and high temperatures may be achieved. This is potentially a 'high grade' energy source as it can be converted to other energy forms relatively efficiently. The second component is the diffuse radiation that is received even when the Sun is obscured by clouds. The diffuse radiation cannot be focused and only provides thermal energy proportional to its natural intensity. It is a 'low grade' energy source. However, photosynthetic organisms can utilize some of the diffuse radiation and produce chemicals in which the energy has been concentrated. The fossilized remains of these chemicals are the source of the coal, oil and natural gas that we now use as 'high grade' energy sources.

4.5 Geochemical accumulation of solar energy

The process of photosynthesis provides plants and ultimately animals with energy by converting solar energy into chemical energy. Any organic remains of dead plants and animals that have been buried in sediments can be thought of as containing trapped solar energy. The majority of these sedimentary organic compounds are dispersed in the rocks at an average concentration of 0.5%, but there are distinct localized deposits (coal, oil, natural gas) where the concentration of organic compounds may be over 90%. These deposits of fossil fuels that are so crucial to the continuation of high-technology societies make up only 0.1% of the total sedimentary organics.

The sites of deposition for the precursors of oil and natural gas were marine, whereas coal, and the associated methane, originates from terrestrial plants. There are still a lot of unknown details about how the deposits were formed. The general stages in the production of fossil fuels were (1) deposition of large quantities of organic matter and of a relatively small quantity of inorganic sediments; (2) anaerobic conditions which prevented oxidation of the organics to carbon dioxide; (3) increasing depth of burial leading to increases in temperature and pressure. In the case of coal, deposition occurred in swampy sites with the organic matter periodically being covered by inorganic sediments, probably brought in by flooding of the area. As the deposit became more deeply buried, the temperature and pressure rose, bringing about changes: peat → lignite → brown coal → bituminous coal → anthracite. During this process the volume of the deposit decreased and there was (1) a loss of hydrogen, nitrogen and oxygen, leading to an increased percentage of carbon; (2) a loss of recognizable plant remains; (3) an increase in hardness; (4) a reduction in volatile matter; (5) an increase in degree of aromatization. **Aromatization** is the formation of six-membered rings of carbon atoms with what were once thought to be alternating single and double bonds (Figure 4.11). In fact the bonding electrons were shared evenly between the carbon atoms. This spreading of electrons is called **delocalization** and gives a more stable structure than that given by localized double bonds. The fusing together of the rings reduces the proportion of hydrogen to very small amounts. Coals are classified by their rank, which indicates how far the change from lignite (low rank) to anthracite (high rank) has proceeded. The same rank of coal can be produced either by a long time at a low temperature or by a shorter period at a higher temperature. Therefore, coals were not produced in only one geological period, though in the UK most coals come from the Carboniferous period.

The reactions bringing about the formation of oil and natural gas are less well known than those bringing about the formation of coal. Oil is mainly composed of linear aliphatic hydrocarbons (few aromatics) usually containing less than 24 carbon atoms and with 50%

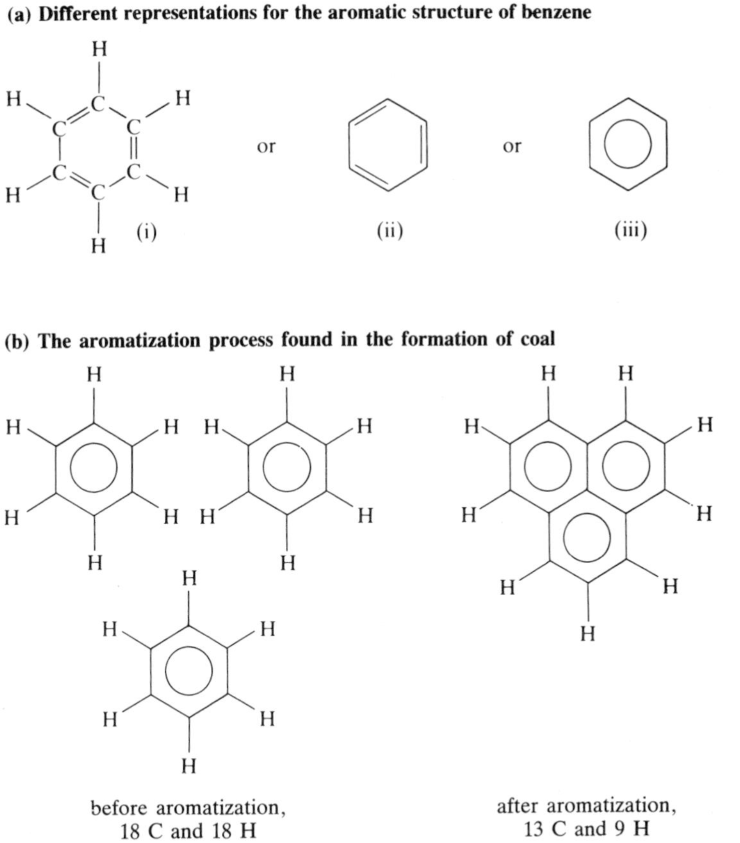

(a) Different representations for the aromatic structure of benzene

or (ii)   or (iii)

(i)

(b) The aromatization process found in the formation of coal

before aromatization,
18 C and 18 H

after aromatization,
13 C and 9 H

Figure 4.11 Aromatic structures. (a) Different representations for the aromatic structure of benzene. (b) The aromatization process found in the formation of coal.

having less than 13 carbons. Oil is fluid at the temperatures of formation and extraction. Natural gas is composed of methane, CH_4, the simplest of the hydrocarbons and the most stable at higher temperatures. It is thought that the majority of the oil and natural gas that has been formed has seeped away and been lost from sediments. The deposits we now exploit, and those we hope to find in the future, result from the oil or natural gas or both having been trapped in a porous rock with an impervious bed above (Figure 4.12).

The burning of all fossil fuels to release energy produces carbon dioxide, water and nitrogen oxides, together with sulphur dioxide in the case of coal and most oil (Eqn 4.22).

$$\underset{\text{natural gas}}{CH_4} \text{ or } \underset{\text{coal and oil}}{C_zH_yS} + \underset{\text{air}}{O_2 + N_2} \longrightarrow H_2O + CO_2 + NO_x + SO_2 \qquad (4.22)$$

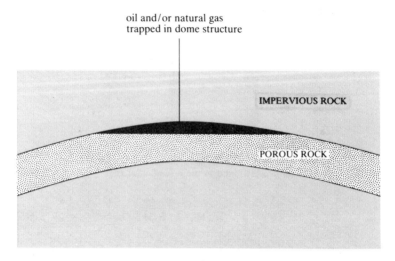

oil and/or natural gas
trapped in dome structure

IMPERVIOUS ROCK

POROUS ROCK

Figure 4.12   The trapping of oil and/or natural gas in sedimentary rocks.

The problems associated with the acidic sulphur and nitrogen oxides are discussed in later chapters. The amount of carbon dioxide released by combustion of fossil fuels has risen from 0.34×10^9 tonnes (99% from solid fuels) in 1860 to 9.2×10^9 tonnes (55% solid fuel, 33% oil, 9% natural gas) in 1960 and 19.5×10^9 tonnes (41% solid fuel, 41% oil, 15% natural gas) in 1986, an average rate of increase of about 3% per year. One of the consequences has been that atmospheric carbon dioxide levels have risen from about 290 ppm in 1860 to 354 ppm in 1990. Carbon dioxide is not toxic at these concentrations and increased amounts in the atmosphere may increase rates of photosynthesis and growth for some plants.

4.6 Greenhouse effect

The heat balance of the Earth will be affected by increased amounts of atmospheric carbon dioxide because the molecules absorb longer-wavelength, infrared electromagnetic radiation, while being transparent to shorter-wavelength, visible and ultraviolet radiation (Figure 4.13). When the Earth reradiates the solar energy that has fallen on its surface the radiation is in the infrared portion of the spectrum. The lower the temperature of the radiation the longer the wavelength of the maximum energy output. This means that the solar radiation is only partially absorbed by carbon dioxide on the 'way in'

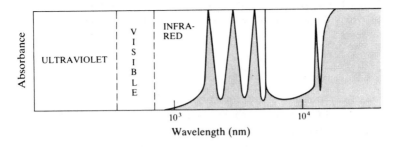

Figure 4.13 The infrared absorption spectrum of carbon dioxide.

(Figure 4.14), but a much higher proportion is absorbed on the 'way out' and trapped in the atmosphere. As the concentration of carbon dioxide rises the amount of radiation trapped will increase and the temperature of the atmosphere will rise. This has been called the **greenhouse effect** because the trapping principle is similar to that operating in a greenhouse, where the glass reduces the loss of infrared radiation. It is now thought that the carbon dioxide concentration will reach twice its pre-industrial level in 70–100 years' time.

Whilst some aspects of the greenhouse effect are known with a high degree of certainty, there is much that is speculative. Trying to piece

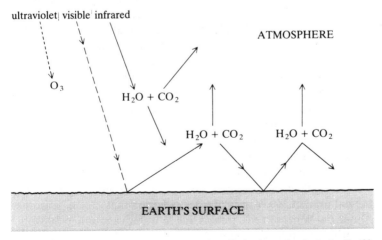

Figure 4.14 The absorption of electromagnetic radiation by molecules in the Earth's atmosphere. The higher the concentration of water (H_2O) and carbon dioxide (CO_2), the greater the proportion of infrared absorbed and then re-emitted to Earth rather than to space until a new equilibrium is reached.

together the likely effects and predict what the consequences will be provides a good illustration of the complexity of environmental systems.

Facts

Gaseous molecules in the atmosphere, such as H_2O, CO_2, CH_4, N_2O, O_3 and CFCs, each absorb characteristic parts of the infrared spectrum (e.g. Figure 4.13, absorption spectrum of carbon dioxide). The amount of infrared absorbed varies from compound to compound and is proportional to the number of molecules of that compound in the atmosphere. The amount of solar radiation entering the Earth's atmosphere is balanced by the amount leaving (Figure 4.14). If there is an increase in infrared-absorbing molecules in the atmosphere, the temperature rises until the radiation emitted again balances the radiation being absorbed. A new steady state is reached. If there was no water vapour or carbon dioxide in the atmosphere the temperature at the Earth's surface would be about 254 K ($-19\,°C$) and life would not exist.

The atmospheric concentrations of CO_2, CH_4, N_2O, O_3 and CFCs are all increasing (Table 4.2). The concentration of water vapour is so variable (0%–5%) that it is not possible to know whether there is an overall increase or not. It should be noted that water vapour is responsible for more infrared radiation absorption than any other atmospheric component. However, it is usually considered that its effect is approximately constant and that changes are driven by other factors. Carbon dioxide and water vapour absorb infrared with wavelengths shorter than 8000 nm and longer than 12 000 nm. This leaves a 'window' of unabsorbed infrared that could escape from the Earth's atmosphere. The recently introduced CFCs and HCFCs are absorbing radiation in the 8000–12 000 nm band. They are, therefore, capable of blocking this 'window' as are ozone, methane and dinitrogen oxide. A doubling of carbon dioxide concentration is calculated to increase infrared absorption by 4 watts per square metre ($4\ W\ m^{-2}$) of the Earth's surface. This extra heat should cause some more water vapour to enter the atmosphere and this would absorb another $2\ W\ m^{-2}$. These increases are relatively small when compared to the present total of $240\ W\ m^{-2}$ of solar radiation that is being absorbed. About $100\ W\ m^{-2}$ of this total is due to water vapour and $50\ W\ m^{-2}$ to carbon dioxide. Doubling the carbon dioxide concentration has such a small effect because a high proportion of the relevant infrared outside the 8000–12 000 nm 'window' has already been absorbed by the carbon dioxide and water already present in the atmosphere.

Table 4.2 Changes in concentration of some greenhouse gases and their average atmospheric lifetimes

Gas	Atmospheric concentration		Annual rate of increase (%)	Atmospheric lifetime (years)
	Pre 1800	1990		
	(ppmv)	(ppmv)		
carbon dioxide	280	353	0.5	150
methane	0.8	1.7	1	10
	(ppbv)	(ppbv)		
carbon monoxide	40	60	0.2	days
dinitrogen oxide	288	310	0.25	150
	(pptv)*	(pptv)		
nitrogen oxides (as nitrogen dioxide)	40	250	0.3	days
non-methane hydrocarbons	100–500	900	?	days
CFC-11	0	280	4	65
CFC-12	0	484	4	130
other Cl compounds	0	~ 1000	1–5	1–400

* Parts per trillion (1×10^{-12}) by volume

Extrapolation and hypothesis

It is possible to extrapolate recent changes in gas concentrations into the future. The validity of the extrapolation will depend upon (a) the precision and accuracy of the measurements made in the past, and (b) the continuation of the processes that produced past increases. If past measurements are incorrect (as appears to have been the case for N_2O, where values were overestimated due to problems with the analytical method), then future projections will also be incorrect. Assumptions about future production and emissions are particularly difficult to make. Whilst the world's population is projected to increase by about 100 million people per year, it is not clear how this will affect greenhouse gas emissions. If living standards rise, use of fossil fuels, and output of carbon dioxide, will rise faster than present trends. Similarly emissions of methane (from paddy fields and from ruminants) and dinitrogen oxide (from fertilizer applications) are dependent upon agricultural productivity. Ozone increases in the troposphere will be affected by the rate of urbanization and increases in motor vehicle usage. CFC production has started to decline, but the known substitutes are also greenhouse gases and the future rate of release of CFCs that are at present trapped in insulating foams and in refrigerators is unknown. To make matters worse, the quantities of materials being transferred between the major sources and sinks are often uncertain. For instance, in the carbon dioxide cycle there appears to be about 1.5–4×10^9 t of carbon dioxide unaccounted for. This represents up to 15% of the extra carbon dioxide that human activities are thought to be releasing to the atmosphere (Table 4.2). Because the extra carbon dioxide that human activities are adding to the atmosphere is only 3–4% of the natural cycle it is not surprising that there is some small discrepancy in the cycle.

The greenhouse gases show a direct effect due to the infrared-absorbing properties of the molecules involved. In addition there can be indirect effects related to the chemical reactions that the gases undergo in the atmosphere. For example, methane shows a direct effect as the methane molecules in the troposphere absorb infrared radiation. Methane is mainly removed from the atmosphere by a series of reactions initiated by the hydroxyl free radicals (Eqn 4.23) that produce carbon dioxide and water, which are greenhouse gases in their own right.

$$CH_4 + HO^* \longrightarrow CH_3^* + H_2O \tag{4.23}$$

$$CH_3^* \text{ series of reactions} \longrightarrow CO_2 + H_2O \tag{4.24}$$

Thus, even when destroyed, the methane leaves behind an 'indirect' effect of radiation absorption by water and carbon dioxide. Hydroxyl radicals are also involved in the removal of carbon monoxide from the troposphere (Eqn 4.25).

$$CO + HO^* + O_2 \longrightarrow CO_2 + \underset{\substack{\text{hydroperoxyl} \\ \text{radical}}}{HCOO^*} \tag{4.25}$$

If the quantities of carbon monoxide and methane are both increasing, there will be competition for the available hydroxyl radicals. Unless there are also new sources of hydroxyl radicals the average lifetimes of these gases will be extended and they will exert their greenhouse effects for a longer period of time.

The hydroxyl radical is involved in the removal mechanisms of the majority of gaseous pollutants. There is great concern that the lack of sufficient hydroxyl radicals will increasingly hinder the ability of the atmosphere to cope with the ever larger quantities of potentially polluting emissions that are entering it. However, some polluting emissions actually produce hydroxyl radicals and so help slightly to alleviate the damage. Tropospheric ozone is a prime example of a hydroxyl radical producer (Eqns 4.26 and 4.27).

$$O_3 + hv \longrightarrow O_2 + O \qquad (4.26)$$

$$H_2O + O \longrightarrow 2HO^* \qquad (4.27)$$

Increased emissions of methane mean that a greater quantity of methane enters the stratosphere and forms hydrogen chloride, HCl, by reacting with chlorine atoms. This reaction breaks the ozone destruction cycle that involves chlorine (see Chapter 2) and helps to preserve the ozone layer. The net effect on temperature changes at the Earth's surface that changes in ozone concentration may have varies with altitude. Ozone in the upper stratosphere absorbs ultraviolet and infrared radiation from the Sun and prevents it reaching the Earth's surface. This warms the stratosphere and cools the troposphere. The ozone in the troposphere absorbs solar ultraviolet and infrared radiation together with some of the infrared emitted from the Earth. This warms the troposphere. The relatively high concentration of ozone at the base of the stratosphere tends to have a proportionately greater absorbing effect on radiation leaving the Earth than on radiation arriving from the Sun. This would imply that the destruction of ozone in the ozone layer would reduce the greenhouse effect.

Any gaseous emission that breaks down to yield a greenhouse gas (e.g. methane, carbon monoxide), or interferes with hydroxyl free radical concentrations (e.g. methane, carbon monoxide, nitrogen oxides, HCFCs), or affects ozone concentrations in the troposphere (e.g. methane, carbon monoxide, nitrogen oxides), or ozone concentrations in the stratosphere (e.g. carbon dioxide, methane, nitrogen oxides, CFCs, HCFCs) will have an indirect effect on the amount of radiation that is absorbed or emitted by the Earth and its atmosphere.

It is possible to attempt to estimate the relative impact that different compounds will have on the absorption of infrared radiation. This relative scale of global warming potentials has each gas compared to carbon dioxide (Table 4.3). The comparison can be on a basis of one molecule to one molecule or on a mass-for-mass basis. The relationship between these two values is determined by the relative molecular masses, e.g. if the per molecule warming potential ratio for $CO_2:CH_4$ is $1:27$, then the per unit mass ratio will be $1:74$ because the respective molecular masses are 44 and 12. The factors that need to be taken into account include:

Table 4.3 Estimates of global warming potentials, over the next 100 years, of greenhouse gases released from anthropogenic sources in 1990

Source	Gas	1990 annual emissions (Tg)	Proportional contribution to global warming potential (%) Time horizon (years) (20)	(100)
fossil fuel combustion	carbon dioxide	19 500	26	44
deforestation		6 500	8	14
	total CO_2	26 000	34	58
agriculture, natural gas	methane	365	39	23
motor vehicles	carbon monoxide	450	4	3
combustion	nitrogen oxides	37	8	3
transportation	(as nitrogen dioxide)	19	4	2
agriculture mainly	dinitrogen oxide	2.8	1	1
various sources	non-methane hydrocarbons	90	3	1
refrigerators	CFC-11	0.29	1	2
air-conditioners	CFC-12	0.42	3	6
foams, solvents aerosols	other Cl compounds	1.1	2	2

(a) the infrared absorption properties of each gas, including the effects of overlapping absorption bands;

(b) the atmospheric lifetime of each gas;

(c) the indirect effects of each gas, as outlined above.

The models usually assume that a small quantity, say 1 kg, of each gas is released into the troposphere. The scaling up of these results to estimate the effects of total present releases, or past or future releases, can then be carried out (Table 4.3). The values are integrated over the required time period to indicate the cumulative effect over that period. Whilst the general trends may be shown, the models are so dependent upon the assumptions being made that they cannot be relied upon for detailed predictions. Table 4.3 indicates that methane is very important immediately after its release, but its short atmospheric lifetime means that its relative importance as a greenhouse gas rapidly declines. It should be noted that, because all of the gases released in one time period are then being removed from the atmosphere, the radiative effect of the gases released in 1990 will be more than halved in 2090.

Hypotheses concerning the climate changes that may arise from increases in greenhouse gases are dependent upon these uncertain extrapolations being introduced into climate models. A general circulation model (GCM) is used to look at possible future global climate trends. This model involves calculating values for various parameters at specific grid points above the Earth's surface. The individual results for each grid point are then combined to give an integrated picture showing changes with time of features such as temperature, rainfall, etc. The GCM is extremely complicated and requires vast amounts of computing power to recalculate the values for each grid point. However, the model can be no better than the assumptions and data that are used in its production. A number of the key features of a GCM are the following.

(a) It has a low resolution (grid points about 200 km apart, though may soon drop to 100 km intervals).
(b) Radiation effects can be described by exact equations, but, to reduce computing demands, approximations are used.
(c) Not much is known about convection effects in the atmosphere, yet these are one of the main ways heat is transferred.
(d) Clouds can either increase absorption of solar radiation because of the water vapour they contain or reflect the solar radiation back into space. Which of these effects would be dominant if cloud cover increased is uncertain.
(e) The interactions between the oceans and the atmosphere with respect to heat transfer, rate of warming and carbon dioxide absorption are uncertain.
(f) There is a lack of knowledge on how land plants will affect the absorption or release of carbon dioxide and also whether methane that is at present trapped in tundra will be released in large quantities if temperatures rise.
(g) If temperatures increase there should be an increase in both evaporation and precipitation of water. If the precipitation occurs over the Antarctic there will be a net removal of water from the oceans to the land because the temperature there will still be low enough for snow to form. How much this will compensate for the expansion of the water in the oceans, as the temperature increases, and for the extra flow of water into the oceans as lower latitude/ altitude glaciers melt, can be only crudely estimated.
(h) There are a number of feedback effects which can increase the rate at which greenhouse gases build up in the atmosphere. Examples include (i) the extra water vapour produced if temperatures rise, (ii) the release of methane that is at present frozen in the soils of

the tundra, (iii) the reduced solubility of carbon dioxide in sea water at higher temperatures, (iv) increased microbial activity in soils breaking down organic matter more rapidly, in what are now temperate climates, to release more carbon dioxide. The concern is that once a global warming event is initiated the rate of change could be self-accelerating.

The results from the various models indicate a variety of temperature rises, rainfall changes, etc. Though these models tend to show the same general trends they are not precise enough to allow detailed planning decisions to be made. In 1992 the 'best' estimates were that a doubling of the pre-industrial carbon dioxide concentration from 280 ppmv to 560 ppmv (in 2060) would give a global average temperature increase of 1.6 K and a 3% increase in precipitation. Changes in the tropics would be smaller, and near the poles larger, than the average figures. As for rises in sea level, figures up to about 40 cm are thought to be most probable.

The movement of warmer climate zones nearer the poles would also move the regions where particular crops grow in a similar manner. For instance, the major grain-growing areas would be shifted northwards in North America and Europe. However, the more northern soils are generally less fertile and thinner. This would suggest that yields would be decreased. Some plants increase their yields in the presence of extra carbon dioxide provided there is sufficient water and nutrients. Most of the main crop plants are described as C_3 plants (an early metabolic product of carbon dioxide fixation is 3-phosphoglyceric acid containing three carbon atoms) and these might give increased yields. This group includes wheat, rice, potatoes, barley and cassava, which between them make up over 50% of global crop production. Another group of plants form oxaloacetic acid, a four-carbon molecule, when they fix carbon dioxide. They do not react so positively to increased concentrations of carbon dioxide. They belong to the C_4 group and examples include maize, cane sugar and sorghum, which have about a 20% share in world food crop production. Most trees are C_3 plants, but again increased stress due to climatic changes may more than wipe out any extra growth resulting from higher carbon dioxide levels.

Some regions will gain in food productivity, others will lose. It seems likely that the climatic changes will not appreciably reduce the productive capacity of the Earth, but they could well disrupt the present imperfect economic and social balance of wealthy and poor nations.

If all the presently estimated reserves of fossil fuels were burnt, the maximum concentration of carbon dioxide in the atmosphere might reach 2000 ppm in 200 or 300 years, giving a temperature rise approaching 6 °C. Of the global coal reserves, about 94% are located in six countries, namely the (former) USSR (45%), the USA (25%), China (14%), Australia (2.5%), (former) West Germany (2.5%) and the UK (1.6%). Coal is just as geographically restricted in occurrence as oil and natural gas, but the reserves are about ten times greater. The actual rate of release of carbon dioxide will be largely decided by the policies of these six countries.

The mean temperature (15 °C) now being experienced in the Northern Hemisphere is higher than the norm (13 °C) over the past 1 million years. During the past 3000 years the mean temperature has usually remained within 1 °C of the present, with warmer centuries occurring about as frequently as cooler ones. The last warming period terminated in 1940, and up to the late 1960s the mean temperature was dropping. This has been followed by erratic behaviour with wider than normal variations making the identification of a definite trend up or down difficult to detect. The rate and magnitude of the carbon dioxide warming that could occur over the next century could be greater than any that have occurred before. The greenhouse effect is a very complicated phenomenon which has highlighted how little is known about the working of biogeochemical cycles and the consequences of their perturbation by human activities. It is unfortunate that arguments about the validity of hypothetical climatic changes and their consequences have obscured the important factor that present trends of resource exploitation are not sustainable. Modification to present fossil-fuel demands is required whether climate or sea level forecasts are correct or not.

5 Nitrogen

Nitrogen Abundance by weight (the relative abundance is given in parentheses): crust, 20 ppm (31); ocean, dissolved N_2, 15.5 ppm (11), soluble compounds, 0.7 ppm (17); atmosphere, 75.53% (1). ppm = mg kg^{-1}.

Nitrogen, in the form of dinitrogen, N_2, makes up 78% by volume (76% by weight) of the Earth's atmosphere. Unlike dioxygen, O_2, the other major component of the atmosphere, dinitrogen has a low chemical reactivity and is not directly available to the majority of living organisms. One consequence of this is that lack of available nitrogen is often a major factor limiting plant growth, and the application of nitrogenous fertilizers produces greatly increased crop yields.

The dinitrogen molecule has three covalent bonds joining the two atoms together, $N \equiv N$. Before a compound can be formed, these bonds must be broken. This process requires 950 kJ mol^{-1} to destroy all three bonds. Despite this very large energy input, a number of microorganisms have been able to develop systems capable of overcoming this energy barrier. As a result, nitrogen cycles in nature at a rate sufficient to allow plants to grow wherever there are adequate water supplies, unless some other exceptional condition prevents growth. The diagrammatic representations of the nitrogen cycle (Figures 5.1 and 5.2) show that there are a number of chemically different, mobile, nitrogen-containing species. The quantitative aspects of many of the transformations in the cycle are not well known and the figures given for the fluxes are often no better than educated guesses.

The relative thermodynamic stability of the various nitrogen species can be determined, and this is shown on the scale headed 'Free energy of formation' in Figure 5.1.

The nitrogen cycle is dominated by reactions involving biological material. All the reactions in the series

and the reverse reactions back to N_2 can be carried out by microorganisms. Natural inorganic reaction mechanisms do not produce

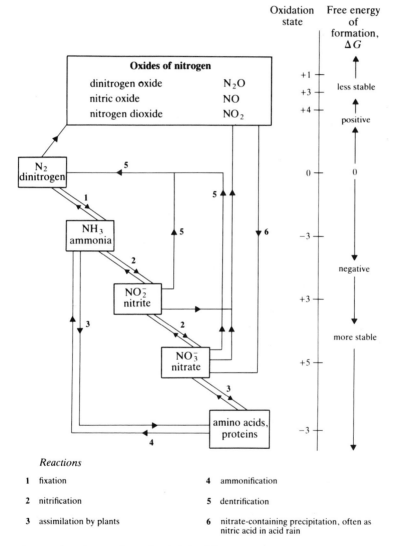

Figure 5.1 Chemical species found in the nitrogen cycle, illustrating changes in oxidation state and relative stability.

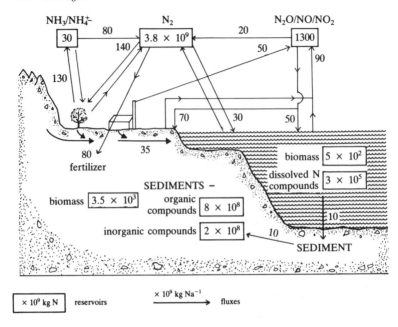

Figure 5.2 The nitrogen cycle, illustrating some of the quantitative aspects of the movement of nitrogen between land, sea and atmosphere.

ammonia from dinitrogen; however, high temperatures (caused by lightning in natural systems, and by furnaces and automobile engines in industrial systems) can cause N_2 and O_2 to combine to form nitrogen oxides. These oxides are gradually removed from the atmosphere as nitrates. This extra supply of nitrate supplements the nitrate produced by micro-organisms.

In 1910–14, Haber and Bosch developed a method for producing ammonia from atmospheric N_2 which has led to the growth of a large-scale nitrogen fertilizer industry. The output of nitrogen fertilizers, about 80×10^6 t a^{-1}, is approaching the estimated fixation of nitrogen by micro-organisms, about 140×10^6 t a^{-1}. The fact that the increase in world food supplies has, overall, kept pace with the increase in population has largely been due to increased use of nitrogen fertilizers.

5.1 Natural transformation processes in the nitrogen cycle

The biologically mediated transformation of dinitrogen into proteins provides organisms with molecules capable of undertaking a wide variety of functions, in tissue structure, enzymes, hormones, etc. The transformation also releases energy (free energies of formation become more negative) that can be utilized by the organism. However, the biogeochemical cycling of nitrogen involves the breakdown of the proteins and the subsequent reversal of the reaction sequence to replace the dinitrogen originally used. Why should micro-organisms apparently use up energy converting the more stable compounds into less stable ones? In practice, of course, a similar problem must occur in all geochemical cycles. If more stable species are produced in one-half of the cycle, the continuation and completion of the cycle demands the subsequent production of less stable species and, consequently, the energy released in one-half of the cycle must be balanced by the absorption of energy in the other half. Examination of the nitrogen cycle (Figure 5.1) shows that the movement of the nitrogen involves (a) changes in oxidation state, (b) changes in energy, (c) reactions with other elements (particularly oxygen, hydrogen and carbon). Closer investigation of these factors should help us to understand why the cycle keeps turning.

The major biological electron-releasing process is the oxidation of carbohydrate to form carbon dioxide (Eqn 5.1).

$$CH_2O + H_2O \longrightarrow CO_2 + 4H^+ + 4e^- \quad \Delta G, -47\,kJ\,mol^{-1}\,e^- \quad (5.1)$$

The fixation of nitrogen requires a source of electrons to bring about the reduction of $N(0)$ to $N(-3)$ (Eqn 5.2).

$$N_2 + 8H^+ + 6e^- \longrightarrow 2NH_4^+ \quad \Delta G, +27\,kJ\,mol^{-1}\,e^- \quad (5.2)$$

This half-reaction tells us that $27\,kJ$ of energy is absorbed for each mole of electrons being taken up by the nitrogen, i.e. the half-reaction will not be favoured, in energy terms, unless it can be combined with another half-reaction that will give the *complete* reaction a negative free-energy change. Combining Equations 5.1 and 5.2 gives Equation 5.3.

$$CH_2O + H_2O + N_2 + 8H^+ + 6e^- \longrightarrow CO_2 + 4H^+ + 4e^- + 2NH_4^+$$
$$\Delta G, (-47 + 27)\,kJ\,mol^{-1}\,e^- \quad (5.1, 5.2)$$

$$3CH_2O + 3H_2O + 2N_2 + 4H^+ \longrightarrow 3CO_2 + 4NH_4^+$$
$$\Delta G, -120\,kJ\,mol^{-1}\,N_2 \quad (5.3)$$

Because this reaction (Eqn 5.3) releases energy, it is thermodynamically favourable and the micro-organisms can derive energy from it. It should be noted in particular that the energy released when the carbohydrate is oxidized more than outweighs the energy required to produce ammonium ions when hydrogen ions, H^+, are the source of hydrogen. By linking the nitrogen cycle to the carbon cycle, the micro-organisms are able to carry out energetically unfavourable half-reactions through the indirect use of solar energy to power the conversion. The apparent paradox of 'impossible' energy changes in the constant revolution of the geochemical cycles is removed as soon as we realize that cycles are interlinked, so that energy is transferred from one cycle to another, and that in addition energy is being constantly added to the terrestrial system, particularly from the Sun.

Though Equation 5.3 expresses the overall process involved in the fixation of N_2, the detailed mechanism is still not completely understood, despite the large amount of research being carried out on the problem. If natural nitrogen fixation could be controlled, so that it replaced the use of industrial nitrogen fertilizers, there would be great savings in energy both in the manufacture and transport of the fertilizer. At the same time there should be less waste, as biological production should match biological requirements. The natural fixation of nitrogen is carried out by micro-organisms, which may be free living or symbiotic, but all of which contain the enzyme nitrogenase. Symbiotic organisms can only survive if they live in association with other organisms. The root nodules of legumes (e.g. peas) contain nitrogen-fixing bacteria, and there are also nitrogen-fixing blue-green algae that are symbiotic with fungi in lichens or with some ferns.

The free-living nitrogen fixers may be either bacteria, some of which live in oxygen-rich (aerobic or oxygenic) conditions whereas others live in oxygen-deficient (anaerobic or anoxic) conditions, or blue-green algae. Nitrogenase is very sensitive to O_2 and those cells that respire aerobically have had to develop various mechanisms to prevent oxidative damage.

The enzyme nitrogenase consists of two large proteins. One of these contains two atoms of molybdenum, 24–32 atoms of iron, 24–30 labile sulphur atoms and has a relative molecular mass of about 220 000, whereas the other protein contains only four atoms of iron, no molybdenum, four labile sulphurs and has a relative molecular mass of about 68 000. The two atoms of molybdenum are essential to the activity of nitrogenase and no fixation will occur if they are absent. The precise mechanism of the reaction is not known, though it has been assumed that N_2 is bound to the molybdenum and reduced at this site by a process such as the following:

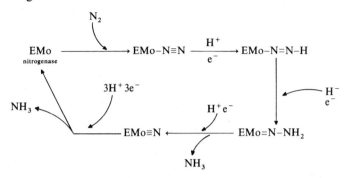

By a process of this type, the organism, instead of taking a single leap over the activation-energy barrier (the energy required to separate the triple-bonded nitrogen molecule into two reactive nitrogen atoms), climbs over the barrier in a number of smaller steps. The energy for the process is transferred to the enzyme via magnesium adenosine triphosphate (MgATP), which is hydrolysed to the diphosphate and free phosphate (Eqn 5.4).

$$MgATP \rightleftharpoons MgADP + PO_4^{3-} + energy \qquad (5.4)$$

The transfer of electrons is favoured by the ability of both iron and molybdenum to exist in two relatively easily interchangeable oxidation states:

$$Fe^{2+} \rightleftharpoons Fe^{3+} + e^- \qquad Mo^{4+} \rightleftharpoons Mo^{5+} + e^-$$

Both the energy and the electrons come ultimately from the oxidation of carbohydrate. This one step in the nitrogen cycle requires the coming together of the nitrogen, carbon, oxygen, hydrogen, sulphur, phosphorus, magnesium, iron and molybdenum cycles, and the essential requirement for the very small quantities of molybdenum highlights the fact that the need for certain critical elements may be out of all proportion to the quantities involved.

The total amount of nitrogen fixed by micro-organisms is difficult to estimate precisely, but is thought to be of the order of $50-150 \times 10^6 \, t \, a^{-1}$. The free-living nitrogen fixers excrete about 80% of their assimilated nitrogen in a plant-available form to produce $1-5 \, kg \, N \, ha^{-1}$ compared to up to $100-300 \, kg \, N \, ha^{-1}$ from leguminous symbiotic fixers. The nitrogen yield varies widely with conditions, being greatest where photosynthesis occurs most readily. It has been estimated that 12% of plant photosynthesis goes to support nitrogen fixation, as three molecules of O_2 are required to oxidize enough glucose to provide the energy to reduce one molecule of N_2.

. steps in the micro-organism-controlled portions of the
.ogen cycle include nitrification, ammonification and denitrification.

Nitrification

This conversion of ammonium ions to nitrate (Eqns 5.5 and 5.6) is
essential for the growth of the majority of plants, as they are able to
absorb nitrate but not ammonia or ammonium.

$$4NH_4^+ + 6O_2 \longrightarrow 4NO_2^- + 8H^+ + 4H_2O \tag{5.5}$$

$$4NO_2^- + 2O_2 \longrightarrow 4NO_3^- \tag{5.6}$$

Unfortunately, nitrate is very soluble in water and easily leached from
soils. Therefore, a build-up of a nitrate reserve in the soil is not possible.

Ammonification

When plants and animals decay, the more complex molecules are
utilized by various organisms and eventually converted into simpler
molecules or ions. The nitrogen-containing compounds eventually
form ammonia or ammonium ions. For example:

$$\underset{\text{urea}}{(NH_2)_2CO} + H_2O \longrightarrow 2NH_3 + CO_2 \tag{5.7}$$

Denitrification

The regeneration of dinitrogen from nitrate occurs under both aerobic
and anaerobic conditions in the soil and oceans. Under anaerobic
conditions organisms can use nitrate to replace dioxygen as an electron
acceptor and as their source of respiratory energy (Eqn 5.8).

$$5CH_2O + 4NO_3^- + 4H^+ \longrightarrow 2N_2 + 5CO_2 + 7H_2O \tag{5.8}$$

The reduction of nitrate does not always form dinitrogen. Appreci-
able quantities of nitrous oxide (dinitrogen oxide), N_2O, may also be
produced. As gaseous N_2O is relatively inert in the troposphere, it is
only slowly removed and is the second most abundant nitrogen
species (0.3 ppm) in the atmosphere. The concentration of N_2O in the
atmosphere is growing at the rate of 0.3% per year.

5.2 Nitrogen oxides in the atmosphere

As well as the biologically mediated reactions, there are a number of
other reactions. Probably the most important of these occur in the

atmosphere and involve the cycling of various oxides of nitrogen. This part of the nitrogen cycle has been the subject of some concern due to the production of increasing quantities of nitrogen oxides by humans.

In addition to the nitrous oxide (N_2O) added to the atmosphere by denitrification reactions, nitric oxide (NO) and nitrogen dioxide (NO_2) are produced at high temperatures (Eqns 5.9 and 5.10): in areas where there are a lot of thunderstorms this may be a significant input source of oxides of nitrogen.

$$N_2 + O_2 \longrightarrow \underset{\text{nitric oxide}}{2NO} \tag{5.9}$$

$$N_2 + 2O_2 \longrightarrow \underset{\text{nitrogen dioxide}}{2NO_2} \tag{5.10}$$

The N_2O gradually passes from the lower atmosphere into the stratosphere where it is rapidly converted into dinitrogen (95%) and nitric oxide (5%). Solar radiation with a wavelength below 250 nm (ultraviolet radiation) has enough energy to break up the N_2O molecules (Eqn 5.11 and 5.12). The photolysis of N_2O can only occur at heights above 20 km because the short-wavelength radiation is absorbed by molecules such as N_2O and O_3 and therefore removed as it passes through the upper atmosphere.

$$N_2O \xrightarrow{h\nu} NO + N \tag{5.11}$$

$$N_2O \xrightarrow{h\nu} N_2 + O \tag{5.12}$$

Further nitric oxide can be formed by the reaction of nitrous oxide with excited (high-energy) oxygen atoms (Eqn 5.13) which are also produced by short-wavelength radiation.

$$N_2O + O \longrightarrow 2NO \tag{5.13}$$

As discussed in Chapter 2, the nitric oxide is able to catalyse the decomposition of ozone which could allow more harmful ultraviolet radiation to reach the surface of the Earth.

Some of the nitric oxide, NO, reacts with either atomic oxygen or ozone to form nitrogen dioxide, NO_2, which then combines with water to give nitric acid, HNO_3 (Eqn 5.14).

$$NO \xrightarrow[\text{or } O_3]{O} NO_2 \xrightarrow{H_2O} HNO_3 \tag{5.14}$$

The nitric acid is then rained or washed out, either as the free acid or combined with ammonia as ammonium nitrate, NH_4NO_3. A number of other reaction pathways involving the oxidation and removal of the nitrogen oxides occur, especially in the lower atmosphere. This is still an area of great uncertainty, mainly because the concentrations of the molecules involved are so low that it is very difficult to trace their fates.

The burning of fossil fuels in furnaces and automobile engines raises the gas temperature sufficiently to allow dinitrogen and dioxygen to combine. In general, the higher the combustion temperature, the greater the quantity of nitrogen oxide. In global terms, anthropogenic emissions from these sources are only about 8% of naturally produced nitrogen oxides. However, in urban areas the relative levels can be elevated several hundred times under certain climatic conditions: whereas the global concentration NO_x is 3×10^{-3} ppm, the concentration in some inner cities is 1–2 ppm. When associated with high hydrocarbon levels and sunlight, these high local concentrations of nitrogen oxides lead to the development of photochemical smogs (Figure 5.3). Photochemical smogs contain a wide variety of organic compounds – such as peroxides (oxidants), aldehydes and ketones (form aerosols), and organo-nitrates (lachrymators) – that are harmful in various degrees.

It is ironic that the reduction in smoke-based smogs in the UK brought about by the Clean Air Acts (see Chapter 6) has provided conditions that favour the formation of photochemical smogs. The formation of photochemical smogs is dependent upon there being high enough concentrations of nitrogen oxides and volatile organic compounds, especially hydrocarbons, together with sufficient sunlight, particularly UV-A. The smoky atmosphere earlier this century reduced the amount of sunlight so that the photochemical reaction sequence was rarely initiated.

Solar radiation provides the energy to break some of the bonds in nitrogen oxides and volatile hydrocarbons. One important sequence of reactions (Eqns 5.17–5.21) is started by the formation of atomic oxygen from nitrogen dioxide (Eqn 5.15).

$$NO_2 + h\nu\,(\lambda < 420\,\text{nm}) \longrightarrow NO + O \qquad (5.15)$$

The reactive atomic oxygen can then produce the following.

(a) Ozone by reacting with dioxygen (Eqn 5.16).
(b) Organic free radicals, that are also very reactive, by reacting with hydrocarbons (Eqn 5.17). It should be noted that about 60% of non-methane volatile organic compounds are from natural sources (trees etc.), with 16% from wood combustion (fuelwood, deforestation) and savanna burning, 7% from gasoline use and 5% from the organic chemicals industry.

(a) Early morning

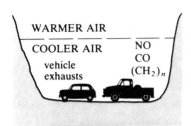

inversion prevents dispersion of
emissions

(b) Later morning

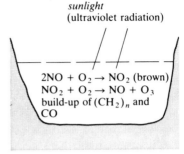

oxidation of NO gives brown colour
to haze – build-up of ozone

(c) Late morning, afternoon

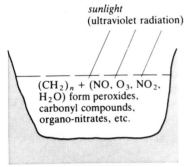

further absorption of solar energy
produces eye irritants and various
other harmful compounds

(d) Night

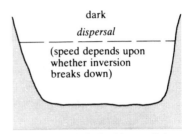

without sunlight (and less traffic)
no further build-up – therefore
partial or complete dispersal is
possible

Figure 5.3 Conditions for smog formation.

(c) Hydroxyl free radicals by reacting with water vapour (Eqn 5.18).
Hydroxyl free radicals are key reactants in many tropospheric
reactions. They are particularly important in controlling the
removal of methane and hence the rate at which the concentration
of this greenhouse gas increases (Chapter 4).

$$O_2 + O + M \longrightarrow O_3 + M \tag{5.16}$$

(M is an energy-absorbing molecule, e.g. N_2)

$$\underset{\text{hydrocarbon}}{RH} + O \longrightarrow \underset{\substack{\text{organic} \\ \text{free radical}}}{R^*} + \text{variety of products} \tag{5.17}$$

$$H_2O + O \longrightarrow \underset{\substack{\text{hydroxyl} \\ \text{free radical}}}{2HO^*} \tag{5.18}$$

The early stages of photochemical smog formation are mainly dependent upon reactions between hydroxyl radicals and hydrocarbons.

$$HO^* + RH \longrightarrow R^* + H_2O \tag{5.19}$$

$$R^* + O_2 \longrightarrow \underset{\substack{\text{organic peroxyl} \\ \text{free radical}}}{ROO^*} \tag{5.20}$$

$$ROO^* + NO \longrightarrow \underset{\substack{\text{organic alkoxyl} \\ \text{free radical}}}{RO^*} + NO_2 \tag{5.21}$$

These reactions do not occur immediately and their occurrence depends upon there being high enough concentrations of the various reactants. As has already been indicated, the subsequent reaction sequence is very complicated but ozone and the products illustrated in Figure 5.3 are typically produced.

If atmospheric conditions tend to be stagnant because of temperature inversions, as is the case in Los Angeles and Mexico City, photochemical smog develops *in situ*. However, as the concentrations of nitrogen oxides and hydrocarbons become generally higher, the presence of elevated levels of tropospheric ozone and the other products of these photochemically induced reactions is also becoming more widespread. The slow rates of reaction mean that the pollutants may appear many miles from the areas where the reactants were released into the atmosphere. For instance, the Grand Canyon in the USA is now being affected by photochemical smog whose source of nitrogen oxides is mainly in Las Vegas. Some of the hydrocarbons are of natural occurrence. Similarly, occasional high concentrations of ozone in rural areas of southern England have been related to polluted air moving north from France and reacting in sunny weather to form ozone.

The increase in the formation of photochemical smogs has followed the increase in use of motor vehicles. These release both nitrogen oxides and hydrocarbons to the atmosphere. Attempts to reduce the photochemically induced pollution have concentrated on altering the exhaust emissions of the vehicles. One method is to fit catalytic convertors to the exhaust system. When the exhaust gases pass through the catalytic convertor system the nitrogen oxides are reduced to dinitrogen (Eqn 5.22) and then the hydrocarbons and any remaining carbon monoxide are oxidized to carbon dioxide and water (Eqns 5.23 and 5.24).

$$2NO + 2CO \longrightarrow N_2 + CO_2 \tag{5.22}$$

$$2CO + O_2 \longrightarrow 2CO_2 \tag{5.23}$$

$$RH + xO_2 \longrightarrow yCO_2 + zH_2O \tag{5.24}$$

A mixture of metals, including the rare metals platinum and palladium, and metal oxides, such as chromium oxide, can be used in the catalysts. The catalytic action is destroyed if the petrol contains lead and this has been the driving force behind the move to unleaded petrol. The catalyst system is most effective when it is hot. This means that on short journeys the expected 90% reduction in nitrogen oxide and hydrocarbon emissions will not occur. The fuel efficiency of cars fitted with catalytic convertors is reduced by about 3% to 10%. Therefore, the reduction in photochemical pollutants by this method will increase emissions of the greenhouse gas carbon dioxide. It appears that the long-term solution to the problems caused by motor vehicles will only be achievable by either changing the propulsion unit or by reducing the number of motor vehicles.

About the same quantity of nitrogen oxides are produced from large stationary sources (power stations and industrial furnaces) as from motor vehicles. Whilst their tall stacks tend to lead to wider dispersal and a lower involvement in photochemical smog formation, they have raised the general background concentration of nitrogen oxides. Methods of control of emissions involve change of burner design, which has already been introduced in a number of power stations, and there is the possibility of treating the flue gases to remove nitrogen oxides if this was required.

There is increasing evidence that in the UK the various emissions from motor vehicles and their photochemically formed products are causing urban health problems, especially respiratory illnesses. Ozone is extremely harmful to plants and it has been suggested that much of the damage that was blamed on 'acid rain' (see Chapter 6) could have been induced by ozone rather than sulphur dioxide/ sulphuric acid/nitric acid. Ozone produces organic free radicals by reacting with the hydrocarbons produced naturally by the plants and it is these that appear to interfere with the normal biochemistry of the plant. However, as nitrogen oxides are also present in the ozone-polluted air, it could well be that these and the associated nitric acid (Eqn 5.14) are responsible for much of the damage.

5.3 Nitrogen fertilizers

There has been a massive increase in the industrial fixation of atmospheric nitrogen over the past 40 years to keep pace with the 20-fold increase in nitrogen fertilizer usage in the UK.

Dinitrogen and dihydrogen combine in the presence of an iron catalyst to produce an equilibrium mixture of reactants and ammonia

(Eqn 5.25).

$$N_{2(g)} + 3H_{2(g)} \underset{80-350 \text{ atm}}{\overset{\underset{400-540\,°C}{\text{Fe catalyst}}}{\rightleftharpoons}} 2NH_{3(g)} \qquad (5.25)$$

The ammonia is removed by condensing it to a liquid and the unused N_2 and H_2 are recycled. Though the conversion rate in any one pass of gases over the catalysts may only be 10%, constant recycling of the reactants means that eventually very high percentage conversions are achieved.

Methane, CH_4, is still used as the source of hydrogen in most countries, even after the rises in the price of natural gas. The methane is mixed with steam and passed over a nickel catalyst to give an equilibrium mixture displaced well to the right of the equation (5.26).

$$CH_{4(g)} + H_2O_{(g)} \underset{\underset{35 \text{ atm}}{750-115\,°C}}{\overset{\text{Ni catalyst}}{\rightleftharpoons}} CO_{(g)} + 3H_{2(g)} \qquad (5.26)$$

It is becoming increasingly more economic to produce ammonia near to the sources of cheap natural gas (such as the Middle East) and then ship it to the user country, rather than to ship the natural gas and produce ammonia near the point of use. Present-day nitrogen fixation plants produce about 1000 t of NH_3 per day and operate at about 80% of the maximum energy efficiency. The use of coal as a feedstock to replace natural gas would increase the energy usage of the plant by 50% and the building costs of the plant would also be about 50% greater.

Nitric acid (HNO_3) is manufactured from ammonia (Eqns 5.27, 5.28 and 5.29) and then combined with more ammonia to form ammonium nitrate (NH_4NO_3), a popular solid fertilizer. (It is also used in explosives.)

$$4NH_3 + 5O_2 \overset{\text{Pt–Rh catalyst}}{\rightleftharpoons} 4NO + 6H_2O \qquad (5.27)$$

$$2NO + O_2 \rightleftharpoons 2NO_2 \qquad (5.28)$$

$$3NO_2 + H_2O \rightleftharpoons 2HNO_3 + NO \qquad (5.29)$$

These reactions are all thermodynamically favourable, but all have kinetic problems that still have to be overcome. The platinum – rhodium catalyst (Eqn 5.27) is only efficient under optimum conditions and various side reactions occur. No catalysts have yet been found for the conversion of nitric oxide, NO, to nitric acid. As a result, very larger reaction towers are required and, as these are not 100% efficient, nitric acid plants are recognizable by the brown–yellow fumes of the mixture of nitrogen dioxide (NO_2) and dinitrogen tetroxide (N_2O_4) that they emit. (NO_2 and N_2O_4 exist in equilibrium with each other, with a higher proportion of NO_2, which is brown, at higher temperatures.)

Over the years there has been an increase in the use of compounds containing higher concentrations of nitrogen, together with higher concentrations of nitrogen in the applied fertilizers. Compounds used include ammonium sulphate (($NH_4)_2 SO_4$), which has 21% N; ammonium nitrate (NH_4NO_3), which has 33.5% N; urea (($NH_2)_2 CO$), which has 45% N; and liquid ammonia (NH_3), which has 82% N. Environmental and economic costs make the correct pattern of application to maximize the nutrients' efficient uptake essential. Increasing the amount of nitrogen applied to a crop gives large increases in yield up to an optimum, but at higher levels the yield may be reduced.

5.4 Nitrate in water supplies

The ploughing up of grassland to grow arable crops and the increasing amounts of nitrogen-based fertilizers that have been used in many countries have led to much higher concentrations of nitrate being found in both ground and surface waters. This nitrate can encourage algal blooms to develop (see Chapter 7) in watercourses. There has been a lot of concern about increased nitrate levels in drinking water. The EC and WHO standards have been set at 50 mg NO_3^- dm^{-1}, but 50–100 mg dm^{-3} may be 'acceptable', though the WHO standard states, 'If the nitrate content is within the acceptable range and the water is otherwise chemically and bacteriologically satisfactory, it may not give rise to trouble, but physicians in the area should be warned of the possibility of infantile methaemoglobinaemia occurring.' It is this risk to children below the age of 6 months that appears to be the major health problem caused by elevated nitrate levels in drinking water. A further potential hazard may be the formation of carcinogenic nitrosamines in the human digestive system by the conversion of nitrate to nitrite and subsequent reaction with amino acids.

Methaemoglobinaemia is the name given to the effects of a reduction in the ability of the haemoglobin in red blood cells to carry oxygen. Normal haemoglobin contains iron(II), Fe^{2+}, which binds reversibly to O_2. Nitrite converts the iron(II) to iron(III), Fe^{3+}, which does not bind to O_2; oxygen can no longer be carried by red blood cells containing these oxidized groups. The haemoglobin in new-born children has a slightly different structure to the haemoglobin of an adult. This foetal haemoglobin (about 60% of total haemoglobin at birth) is more susceptible to oxidation. In the first few months after birth the proportion of foetal haemoglobin declines, as does the susceptibility to methaemoglobinaemia. In addition, the conditions in the gastro-intestinal tracts of very young children are more favourable to nitrate-reducing bacteria than the conditions in the digestive tracts of older humans. Therefore more of the ingested nitrate is reduced to nitrite, which then reacts with the haemoglobin. Cattle also have digestive systems capable of supporting nitrate-reducing bacteria and they too can suffer from methaemoglobinaemia if nitrate levels are high in the water they drink.

5.5 Balanced diets and food production

Though proteins, carbohydrates and fats are the major dietary requirements for humans, many other compounds are needed to give a healthy body. What constitutes a complete balanced diet is not precisely known, but the major features of such a diet are recognized. The WHO has produced recommended daily allowances (RDAs) for energy, protein, vitamins and minerals for reasonably healthy individuals. The RDAs vary with age and sex, but because of natural individual variations and the inadequacy of the data on which the balances are based, the recommended intakes may be insufficient for some people and overgenerous for others.

Nitrogen is only taken up by plants in certain forms, mainly nitrate, and then used to produce proteins. Animals cannot produce proteins from simple ions, but must ingest proteins and use these as a source of amino acids to synthesize the proteins they require. Therefore, proteins are an essential requirement of any animal's diet. Some animals obtain the required proteins from plants (**herbivores**), others feed off other animals (**carnivores**). Humans belong to the group that utilizes both plants and animals (**omnivores**). Proteins are not required as a major energy source for human beings: we get energy from carbohydrates and fats.

Not only must the diet contain enough protein – about 0.8 g protein per kg body weight – but these proteins must be of the correct type. Humans are unable to synthesize nine essential amino acids (Table 5.1) and these must be supplied in the proteins eaten. Different foods contain different quantities of the essential amino acids. Some protein sources, e.g. eggs, fish, meat, are described as 'high quality' because they contain high quantities of all the essential amino acids, whereas most vegetable protein is 'low quality' because it is deficient

Table 5.1 The amounts of essential amino acids in some foods (mg amino acid per g food)

Name	Abbreviation	Beans	Maize	Rice	Cows' milk	Beef	Hens' eggs
isoleucine	Ile	37	30	36	47	53	54
leucine	Leu	74	116	79	95	82	86
lysine	Lys	94	30	42	78	87	70
methionine and cystine	Met + Cys	47	43	40	33	38	57
phenylalanine and tyrosine	Phe + Tyr	68	84	99	102	75	93
threonine	Thr	56	40	33	44	43	47
tryptophan	Trp	12	8	14	14	12	17
valine	Val	39	51	57	64	55	66

in one or more of the essential amino acids (Table 5.1). The deficiency of just one amino acid reduces the degree of utilization of all the rest, because the proteins that are synthesized from them contain fixed proportions of each amino acid. For instance, an adult requires 1 part of phenylalanine to 0.73 parts of lysine. In rice, there are only 0.42 parts of lysine to 1 part of phenylalanine. As there is not sufficient lysine to allow full utilization of the phenylalanine, rice is relatively deficient in lysine. Vegetarian diets must contain complementary protein sources. For example, by combining legumes (e.g. beans and peas) that are higher in lysine and lower in methionine, with cereals (e.g. wheat, rice and maize) that are lower in lysine and higher in methionine, a balanced protein diet can be achieved.

In developing countries, the majority of the population relies on a few crops for its staple diet. Amino-acid deficiencies may be widespread if the correct balance of cereals, legumes and vegetables is not achieved. Worldwide, the increases in population have been balanced by increases in food production so that the average energy intake has remained steady. The increased production has been achieved by a particularly large increase in cereal yields and resulting diets have become less varied. This dependency on a few species is highlighted by the fact that five crops (wheat, rice, maize, potato and barley) appear to contribute a greater tonnage to world food stocks than all other food crops added together. Though a worldwide balance is being achieved, there is insufficient transfer from overproducing to underproducing areas. Malnutrition and famine will remain major problems until problems of transport and wastage are overcome.

The use of animal protein to give better balanced diets in the developing countries suffers from a number of problems:

(a) although many animals will eat cellulose-based plant material, which humans cannot digest, there are few parts of the world where this can be grown in large enough quantities on land that could not be used for growing directly usable food crops;

(b) animal products deteriorate very rapidly and their effective utilization depends upon preservation methods that are often energy expensive and demand relatively high technology;

(c) many countries have religious and cultural restrictions that prevent either the killing or consumption of animals;

(d) the conversion of vegetable protein to animal protein is a very inefficient process, ranging from about 4% efficiency in beef production to about 20% efficiency in poultry production.

It would seem that amino-acid deficiencies will be best overcome by varied vegetable-protein-based diets, and, towards this end, changes in the genetic stock of non-cereal crops to increase yield and reduce disease susceptibility are being actively researched.

Smaller quantities of vitamins and mineral elements are required in addition to the major components of the diet, carbohydrates, fats and proteins, which provide energy and build up or renew body tissue. The **mineral elements** include all the elements apart from carbon, hydrogen, oxygen and nitrogen that are essential for the correct functioning of the body. The functions include formation of the skeleton, maintenance of the electrolytic balance in body fluids, electrochemical stimulus of nerves and muscle, activation and structure of enzymes, and oxygen transport. Details of the function of individual elements will be discussed in the relevant chapters. Vitamins are organic compounds of varying degrees of complexity whose RDAs vary from 60 mg for vitamin C to 6 μg for vitamin B_{12}.

Vitamins were named before their chemical constitutions were known; letters were used to distinguish the various extracts that were found to be essential for life (Latin *vita* = life). As extraction and separation techniques improved, some of the vitamin extracts were found to contain a number of active ingredients which were distinguished by the use of added numbers, e.g. B_1, B_2. When the chemical structures had been determined, trivial names were given (Table 5.2). Apart from vitamin D, which can be synthesized in the skin when it is exposed to sunlight (and therefore may not strictly be a vitamin), the vitamins must be eaten as part of the diet. The vitamins may be classified as water soluble or fat soluble. The water-soluble vitamins are easily eliminated from the body, in the urine, and removed from foodstuffs, if cooked in water. They must be present in the daily diet if deficiencies are to be avoided. The fat-soluble vitamins are more

Table 5.2 Vitamins, their recommended daily allowances (RDAs), and possible effects of deficiency in human diets

Vitamin	RDA Male	RDA Female	Deficiency effects
Water-soluble			
B group			
thiamine (B$_1$)	1–1.5 mg	0.7–1.1 mg	beriberi
riboflavin (B$_2$)	1–7 mg	1.3–1.8 mg	cracked skin, eye lesions
niacin	18 mg	15–21 mg	pellagra
pyridoxine (B$_6$)	2 mg	2 mg	anaemia, dermatitis
cyanocobalamin (B$_{12}$)	0.003 mg		pernicious anaemia
pantothenic acid	10 ng		neuromotor (rare)
folic acid	c. 0.4 ng		anaemia
biotin	0.3 ng		dermatitis
ascorbic acid (C)	30 mg	30–60 mg	scurvy
Fat-soluble			
retinol (A)	0.75 mg	0.75–1.2 mg	eye disease, skin lesions
cholecalciferol (D)	2.5 ng	2.5–10 ng	rickets
tocopherols (E)	15 mg		not known
phylloquinone (K)	0.03 mg		internal haemorrhages

stable, being less easily destroyed by heat; any vitamins in excess of daily requirements are stored in the liver. This storage process means that there is not the same requirement for regular daily supplies. However, these vitamins are also quite toxic, unlike the water-soluble ones, and excessive intakes, particularly of vitamins A and D, can lead to undesirable effects.

6 Sulphur

Sulphur Abundance by weight (the relative abundance is given in parentheses): Earth 1.8–2.9% (5 or 7); crust 260 ppm (16); ocean, 905 ppm (6); atmosphere, 0.6 ppb (15). ppm = mg kg^{-1}; ppb = μg kg^{-1}.

6.1 The sulphur cycle

The range of sulphur concentrations given for the Earth indicates the uncertainty of our knowledge of the inner mantle and core where most of the sulphur is concentrated.

The cycling of sulphur at the Earth's surface has been greatly increased since the start of the Industrial Revolution by the demand for fuel, metals and fertilizers. Despite the great deal of study the sulphur cycle has received in the past few years, there is still some

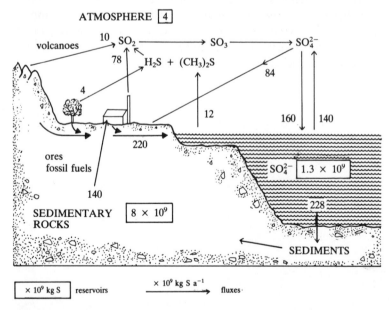

Figure 6.1 Some quantitative aspects of the sulphur cycle.

uncertainty about many of the fluxes (Figure 6.1). In the case of anthropogenic emissions to the atmosphere, knowledge of the extraction and usage of fossil fuels and sulphide ores gives a good indication of gaseous sulphur oxide outputs. It is noticeable that the majority of the anthropogenic effects lead to an increase in oxidation state for the sulphur (Figure 6.2), whereas the biologically mediated natural cycle of sulphur has a large component in which the oxidation state is reduced. The cycles of sulphur and nitrogen have a number of similarities which will be discussed in the following section, but one most important difference is that the major reservoir for nitrogen is the atmosphere, whereas the major available reservoir for sulphur is the crust, with only a small, but potentially damaging, proportion in the atmosphere.

Micro-organisms and the sulphur cycle

There is a strong analogy between sulphur and nitrogen in the way that micro-organisms influence their biogeochemical cycling. Each element tends to be present in living organisms in its most reduced form, i.e. nitrogen (-3) as amino groups, $-NH_2$, and sulphur (-2) as hydrosulphide groups, $-SH$. Sulphur is an important secondary constituent of amino acids and proteins. The ability of this sulphur to form sulphur–sulphur bonds allows cross-linking in proteins by so-called disulphide linkages (Figure 6.3). There can be both intermolecular bonding linking together adjacent proteins and so developing large-scale structures (e.g. hair and nails) and intra-molecular bonding, giving specific spatial characteristics to the protein (e.g. enzymes). When organic sulphur compounds are decomposed by bacteria, the initial excreted sulphur product is generally hydrogen sulphide, H_2S (Eqn 6.1), in the same way that organic nitrogen compounds yield the hydrogen-containing products ammonia, NH_3, or ammonium ions, NH_4^+.

$$R-SH \xrightarrow{\text{bacteria}} H_2S + RH \qquad (6.1)$$

organic
hydrosulphide   organic
compound

Many marine phytoplankton produce compounds that break down to produce dimethyl sulphide, $(CH_3)_2S$. It is thought that dimethyl sulphide is the major biogenically produced sulphur compound released from oceans. Dihydrogen sulphide, H_2S, is more important in terrestrial and salt-marsh environments. Dimethyl sulphide, like dihydrogen sulphide, is rapidly oxidized to form sulphur dioxide and, ultimately, sulphate (Eqn 6.2). Some micro-organisms in muds

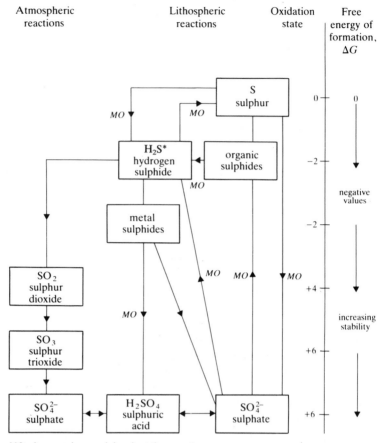

MO micro-organisms can bring about these reactions
* marine phytoplankton often release dimethyl sulphide, (CH₃)₂S

Figure 6.2 Chemical species found in the sulphur cycle, illustrating changes in oxidation state and relative stability.

produce elemental sulphur, a yellow solid, which is comparable with the production of dinitrogen gas from nitrogen compounds.

The oxidation of the reduced forms of sulphur by dioxygen (Eqn 6.2) may occur either through the action of micro-organisms in the soil, the sediment and the water column, or without biological control, particularly in the atmosphere.

$$H_2S \longrightarrow (S) \xrightarrow{O_2} SO_2 \xrightarrow{O_2} SO_3 \longrightarrow SO_4^{2-} \quad (6.2)$$

hydrogen sulphide   (sulphur: may not be formed)   sulphur dioxide   sulphur trioxide   sulphate

(a) The formation of a disulphide linkage by the combination of two cysteine groups on adjacent segments of protein chains – the reaction occurs under reducing conditions

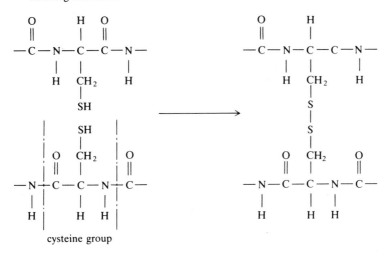

cysteine group

(b) Inter-molecular bonding joining two protein molecules together

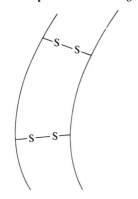

(c) Intra-molecular bonding holding a protein chain in a particular shape

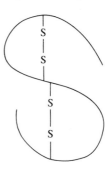

Figure 6.3 The formation of sulphur–sulphur bonds (disulphide linkages) in proteins.

The half-reaction for the reduction of sulphate (Eqn 6.3) is energetically unfavourable.

$$SO_4^{2-} + 9H^+ + 8e^- \longrightarrow HS^- + 4H_2O \quad \Delta G, + 21.4 \, kJ \, mol^{-1} e^- \quad (6.3)$$

To carry out this half-reaction the micro-organism combines it with a half-reaction such as the oxidation of carbohydrate that will give an overall negative free-energy change, leading to sulphate reduction (Eqn 6.4).

$$SO_4^{2-} + H^+ + 2CH_2O \longrightarrow HS^- + 2H_2O + 2CO_2 \quad \Delta G, -25.6\,kJ\,mol^{-1}e^-$$

$$(6.4)$$

The sulphate is acting as an electron acceptor and the oxidation of the carbohydrate is able to proceed. The energy released to the organism is only about one-fifth of that which is obtainable when dioxygen acts as the electron acceptor in aerobic respiration (Eqn 6.5). Nitrate can also act as an electron acceptor (Eqn 6.6). This reaction releases three times as much energy as the reduction of sulphate (Eqn 6.4), but sulphate reduction does produce slightly more energy than is obtained by the methane-producing fermentation reaction (Eqn 6.7).

$$CH_2O + O_2 \longrightarrow CO_2 + H_2O \quad \Delta G, -125.5\,kJ\,mol^{-1}e^- \quad (6.5)$$

$$2CH_2O + NO_3^- + 2H^+ \longrightarrow 2CO_2 + H_2O + NH_4^+ \quad \Delta G, -82.2\,kJ\,mol^{-1}e^-$$

$$(6.6)$$

$$2CH_2O \longrightarrow CH_4 + CO_2 \quad \Delta G, -23.5\,kJ\,mol^{-1}e^- \quad (6.7)$$

The consequences of this order of energy production are that if all four mechanisms are possible, then organisms that gain their energy by aerobic respiration will be favoured (Figure 6.4). The greater quantity of energy that the aerobic respirators obtain from a given carbohydrate source will allow them to develop at the expense of micro-organisms depending upon less efficient energy-releasing processes. When the dioxygen supply becomes depleted and the conditions become anaerobic, the nitrate reducers will be most favoured. The nitrate concentration is much lower than the sulphate concentration under marine conditions and the nitrate is soon depleted, allowing the sulphate-reducing micro-organisms to become dominant. The formation of hydrogen sulphide (H_2S) is a characteristic feature of anaerobic marine sediments due to the high levels of sulphate, as compared with nitrate, in sea water. In soils, where nitrate levels are often higher than sulphate levels, ammonia is produced under anaerobic conditions. When sulphate concentrations are very low, methane bacteria become the dominant form. Examples include (a) fresh waters, in which sulphate is naturally at a low level; (b) marine muds, in which the sulphate has already been converted to sulphide; (c) sewage sludge, in which both nitrate and sulphate are at low levels. In oceanic surface waters dimethyl sulphide is formed much more commonly than H_2S because of the presence of the compound dimethyl sulphonopropionate produced by some species of phytoplankton.

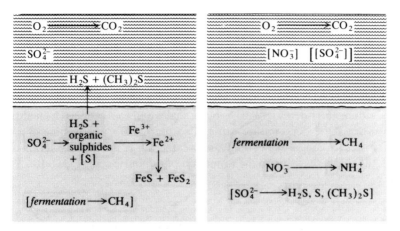

(a) Marine conditions

$O_2 \longrightarrow CO_2$

SO_4^{2-}

$H_2S + (CH_3)_2S$

$SO_4^{2-} \rightarrow$ H_2S + organic sulphides + [S] $\quad$ Fe^{3+} $\longrightarrow$ Fe^{2+}

$\downarrow$

FeS + FeS$_2$

[*fermentation* $\longrightarrow$ CH$_4$]

(b) Fresh-water conditions

$O_2 \longrightarrow CO_2$

$[NO_3^-] \quad [[SO_4^{2-}]]$

fermentation $\longrightarrow$ CH$_4$

$NO_3^- \longrightarrow$ NH$_4^+$

$[SO_4^{2-} \longrightarrow$ H$_2$S, S, (CH$_3$)$_2$S]

for marine and fresh-water systems, the surface waters are generally aerobic with a plentiful supply of dioxygen, but at depth or in the sediments conditions become anaerobic

(c) Soils

$O_2 \longrightarrow CO_2$

$NO_3^- \longrightarrow$ NH$_4^+$

fermentation $\longrightarrow$ CH$_4$

anaerobic conditions result when the soil is saturated with water or an impermeable layer forms preventing access by dioxygen – fermentation reactions are particularly common in buried refuse tips

Figure 6.4 Electron acceptors found under different natural environmental conditions.

Dihydrogen sulphide is associated more with anaerobic sediments like those that are found in swamps and on sea-floors.

The hydrogen sulphide that is produced may be released as a gas to the atmosphere, where it is oxidized (Eqn 6.2) (as is dimethyl sulphide), or it may react with metal ions in the sediments or water column to form insoluble sulphides. The later transition metals and those metals which come after the transition metals in the periodic table are especially likely to form insoluble sulphides. Iron, because it is present in relatively large quantities, forms the major sulphide mineral reservoir as troilite, FeS (Eqn 6.8), and as iron pyrites, FeS$_2$

(Eqn 6.9).

$$2Fe(OH)_3 + 3H_2S \longrightarrow 2FeS + S + 6H_2O \qquad (6.8)$$
iron (III)   troilite
hydroxide

$$FeS + S \longrightarrow FeS_2 \qquad (6.9)$$
pyrites

The black colour of many sediments is partially due to the presence of iron sulphides as well as organic matter. The association of sulphide minerals, particularly iron pyrites, with rich organic marine sedimentary deposits such as black shales and coal can lead to acidity when these deposits are exposed to the Earth's atmosphere. Such exposure can be due to uplift and weathering or mining. Oxidation of the sulphides leads to the formation of sulphuric acid (Eqn 6.10).

$$2FeS_2 + 2H_2O + 7O_2 \longrightarrow 2FeSO_4 + 2H_2SO_4 \qquad (6.10)$$
pyrites   iron (II)   sulphuric
                                 sulphate   acid

This redox reaction can occur rapidly in the presence of water and dioxygen, especially when micro-organisms are involved. The acid produced may be neutralized as it passes through rocks and soil in the ground water especially if it is formed slowly by natural weathering processes. Because of the greater solubility of many metals in acid solutions, extensive leaching occurs. Many black shales contain high concentrations of potentially toxic elements and the increased mobilization of these metals in the acidic waters may cause elevated levels to appear in plants and aquatic biota. The problem becomes much more serious in metalliferous mine-waste dumps that can have some toxic metals present in parts per hundred, rather than the parts per million of normal rocks and soils. The mining of metal sulphide ores and coal leads to the exposure of large quantities of sulphides and hence to the production of much larger quantities of acid in localized areas than does the weathering of shales. The resultant large addition of acid, sometimes enriched in toxic metals, can lead to major pollution problems in the areas surrounding the mining operation.

6.2 The sulphur dioxide problem

Sulphur dioxide (SO_2) is introduced into the atmosphere by both natural and human activities. The amount of sulphur dioxide produced by anthropogenic sources in 1990 was estimated to be 156 million tonnes compared to 52 million tonnes from natural sources; but in the Northern Hemisphere, anthropogenic sulphur dioxide totalled 140 million tonnes whereas natural sources totalled only 30 million tonnes. This difference between Northern and Southern

Hemispheres reflects the higher proportions of land mass and so-called developed countries which use large quantities of fossil fuels.

The sulphur dioxide in the atmosphere is oxidized by a wide variety of mechanisms that include gas–gas interactions called **homogeneous oxidation** (as there are no phase differences; Eqn 6.11) and reactions occurring in solution, with or without the presence of catalysing agents (called **heterogeneous oxidation** if there are phase differences between reactants; Eqn 6.12a and b). Manganese appears to be the most effective catalyst.

$$HO^{\cdot} + SO_2 \xrightarrow{\hspace{1.5cm}} HSO_3^{\cdot} \xrightarrow{O_2} HSO_5^{\cdot} \xrightarrow{NO} HSO_4^{\cdot}$$

$$+ NO_2 \xrightarrow{H_2O} H_2SO_4 + HNO_3 \qquad (6.11)$$

$$SO_{2(aq)} + O_{3(g)} \xrightarrow{\hspace{1.5cm}} SO_{3(aq)} + O_{2(g)} \xrightarrow{\hspace{1.5cm}} H_2SO_4 \qquad (6.12a)$$

alternatively

$$SO_{2(aq)} + H_2O_{2(aq)} \xrightarrow{\hspace{1.5cm}} SO_{3(aq)} + H_2O \xrightarrow{\hspace{1.5cm}} H_2SO_4 \qquad (6.12b)$$

Reaction sequence 6.11 involves a number of **free radicals** ($HO^{\cdot}$, $HSO_3^{\cdot}$, $HSO_5^{\cdot}$, $HSO_4^{\cdot}$) all of which are reactive species containing a single unpaired electron, indicated by the dot in the symbol. Free radicals can be formed in the atmosphere by a large number of reactions, usually involving photolysis, in which a covalent bond is split by the absorption of solar radiation. The ultraviolet part of the solar spectrum is the portion normally involved (Eqn 6.13).

$$\underset{\text{ozone}}{O_3} \xrightarrow[\substack{\text{wavelength} \\ 310\,\text{nm}}]{h\nu} O_2 + \underset{\substack{\text{excited} \\ \text{oxygen} \\ \text{atom}}}{O^*} \xrightarrow{H_2O} \underset{\substack{\text{hydroxyl} \\ \text{free} \\ \text{radical}}}{2HO^{\cdot}} \qquad (6.13)$$

Depending upon the amount of moisture in the atmosphere, 20–80% of the sulphur dioxide emitted into the air is oxidized to sulphate whilst the remainder is removed by dry deposition (Figure 6.5). Sulphate has a relatively small depositional velocity and most of it is removed by wet deposition (Figure 6.5). The sulphur dioxide and sulphate mixture has an average lifetime of 2 to 6 days in the atmosphere during which time it may travel up to 4000 km from its source. **Dry deposition**, despite its name, involves sorption of the sulphur compounds mainly on moist surfaces such as vegetation, wet buildings, soil, and water bodies. **Wet deposition** involves removal by rain or snow and is also called 'precipitation scavenging'. Material entering clouds is **rained out** whereas material below clouds is **washed out**. Most sulphur oxides entering a rain belt will be removed within 200 km.

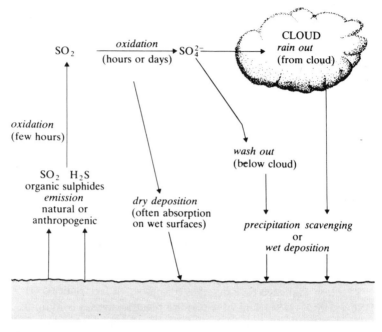

Figure 6.5 Mechanisms by which sulphur dioxide and its oxidation products are removed from the atmosphere.

The actual mechanism of dispersion of sulphur dioxide from point sources such as power stations or smelters is very complicated, being dependent upon climatic conditions, topography, chimney height and design. In general the higher the chimney the more widely will the sulphur dioxide be dispersed, but lack of atmospheric turbulence can lead to a concentrated plume travelling hundreds of kilometres before dispersing. Very low chimneys, as on houses, are especially ineffective in dispersing sulphur compounds. In the UK, the average concentration of sulphur dioxide in the atmosphere is about $33\,\mu\mathrm{g\,m^{-3}}$, of which about 90% is from anthropogenic sources. However, this average value hides wide variations between different areas of the country and between different periods. The range of daily average readings is typically 50 times greater than the annual range. Five-minute averages and 1 hour averages may be respectively ten times and five times greater than the daily average. The short-term variations are mainly due to changes in the meteorological conditions. The spatial distribution of sulphur dioxide concentrations closely follows the spatial distribution of low and medium-height chimneys. The highest sulphur dioxide concentrations are in the larger towns and the much lower levels in rural areas may be only 10–20% of the central urban values. The average urban values for the UK were $188 \mu g SO_2 m^{-3}$ in 1958, $144 \mu g SO_2 m^{-3}$ in 1970, and $73 \mu g SO_2 m^{-3}$ in 1977. This reduction illustrates very clearly the success of the enforcement of the Clean Air Acts (1956 and 1968) and the conversion of houses to the convenient heat sources of natural gas and electricity. In the same period, the emission of SO_2 from tall chimneys, largely associated with electricity power stations, has increased from 1.4 million tonnes to 3 million tonnes, but the emissions from low chimneys have decreased from 1.7 million tonnes to 0.6 million tonnes.

The extent of acidic rainfall appears to be growing because the use of taller chimneys causes wider dispersion of the sulphur and nitrogen oxides. Localized very high levels have been reduced, but delocalized higher levels have been increased. Rain normally has a pH of 5.6, when in equilibrium with the carbon dioxide in the atmosphere. If the pH drops below 5.6, the rain is described as being acidic. The sensitivity of an ecosystem to acid rain depends upon the quantity of acid deposited, upon the ability of the rock, soil and water systems to neutralize the acid, and upon the resistance of the living organisms to any consequent changes.

Rocks may be classified into four types based on their ability to neutralize acids (Table 6.1). The essential features to note are that the rocks that are most resistant to normal chemical weathering also have the lowest ability to neutralize acids; and that the greater the proportion of calcium and magnesium carbonates in the rocks, the greater the neutralizing effect. Soil composition is partially a function of the underlying rocks, but the degree of weathering and the amount of organic matter can greatly modify the soil's properties.

Cation-exchange capacity has been used to define the sensitivity of soils to acid precipitation (Table 6.2). Clay minerals and organic matter contain sites at which hydrogen ions can be attached to the solid and at which metal ions can be released into the solution (Eqn 6.14). This cation exchange lowers the concentration of hydrogen ions in the soil water.

$$(\text{Clay/organic matter}) - M^+ + \underset{\substack{\text{free} \\ \text{hydrogen} \\ \text{iron}}}{H^+} \longrightarrow (\text{clay/organic matter}) - \underset{\substack{\text{bound} \\ \text{hydrogen} \\ \text{ion}}}{H^+} + M^+$$

$$(6.14)$$

The cation-exchange capacity of the soil is limited and the presence of free carbonate minerals is of overwhelming importance. Other exchange reactions are also possible (Chapter 12).

Table 6.1   Classification of rock formations according to their ability to neutralize acid (Reprinted with permission from Glass, N. R. *et al.* 1982. Effects of acid precipitation. *Environ. Sci. Technol.* **16**, 162A–8A. Copyright © 1982 American Chemical Society)

Type No.	Description and examples
I	Low or no buffering capacity, overlying waters very sensitive to acidification; e.g. granite, syenite, granite gneisses, quartz sandstones or equivalents
II	Medium to low buffering capacity, acidification restricted to first- and second-order streams and small lakes; e.g. sandstones, shales, conglomerates, high-grade metamorphic rocks, intermediate igneous rocks, calc-silicate gneisses
III	High to medium buffering capacity, no acidification except in cases of overland runoff in areas of frozen ground; e.g. slightly calcareous, low-grade metamorphic rocks, intermediate to basic, ultrabasic and glassy volcanic rocks
IV	'Infinite' buffering capacity, no acid precipitation of any kind; e.g. highly fossiliferous sediments or metamorphic equivalents, limestones, dolomites

The major factor controlling the ability of natural water systems to absorb acid inputs without major changes in pH is the alkalinity (Chapter 4). The lower the alkalinity, the greater the sensitivity to acid rainfall. It is clear that the most sensitive members of the three systems (rock, soil, and water) will usually occur in the same area (a) because

Table 6.2   Soil sensitivity to acid precipitation, based on the chemical characteristics of the top 25 cm of soil (Reprinted with permission from Glass, N. R. *et al.* 1982. Effects of acid precipitation. *Environ. Sci. Technol.* **16**, 162A–8A. Copyright © 1982 American Chemical Society)

Sensitivity	Cation-exchange capacity (meq* per 100 g)	Other relevant conditions
non-sensitive (a)	any value	free carbonates present, or subject to frequent flooding
(b)	> 15.4	none
slightly sensitive	values from 6.2 to 15.4	free carbonate absent; not subject to frequent flooding
sensitive	< 6.2	free carbonate absent; not subject to frequent flooding

*Milli-equivalent – see page 187

slow-weathering rocks form thin soils containing few clay minerals, and generally the nutrient content is low, (b) because the poor soils support less vegetation which means in turn that the soils have low organic matter contents, (c) because as the runoff water has picked up little calcium and magnesium carbonate from the soils, the streams, rivers and lakes all have low alkalinities.

The effects of acid rain may be directly related to the acidity of the rainfall, as seen in the dissolution of limestone statues, buildings and rock outcrops. In other cases, the effects may be indirect, as when the acid causes the release of aluminium in soils which then reaches levels at which it is toxic to trees. There are also possible beneficial effects to be taken into account, such as the increase in levels of available sulphur and nitrogen which may reduce or remove deficiencies in these nutrients. Distinguishing the indirect effects in complex ecosystems is extremely difficult, particularly as the effects may only become apparent after several years. For instance, there was no change in the composition of water collected 20 cm below the surface of a soil in a forest until 2 years after the start of the addition of simulated acid rain, pH 4. The concentration of hydrogen, sulphate, calcium and magnesium ions then all rose together.

The effects of acid rain and of sulphur dioxide emissions are very varied and a number of them are briefly discussed below.

Human health

Levels of SO_2 above $500 \,\mu g \, m^{-3}$ (24 hour average) affect asthma and bronchitis sufferers if levels of smoke are about $250 \,\mu g \, m^{-3}$. Increases in sudden deaths and hospital admissions are noted at about $750 \,\mu g \, m^{-3}$ for both SO_2 and smoke. The infamous London fog of 5–8 December 1952 led to 4000 deaths with SO_2 and smoke levels of about $4000 \,\mu g \, m^{-3}$. The relative importance of sulphur dioxide appears to be much less than that of smoke, but each enhances the effect of the other. The identification of the causes of particular effects on health is usually a very long, time-consuming and expensive enterprise, because of the lack of suitable data bases. In the case of SO_2 and smoke measurements, should one compare average values for 5 minute, 1 hour, 24 hour, 1 week, 1 month or 1 year periods? Where should the sampling stations be situated relative to the population being studied? It seems that for many health effects it is the incidence of occasional high levels of smoke and SO_2 that is important rather than the long-term average.

In 1961, before the 1956 Clean Air Act had become effective, smoke and sulphur dioxide levels were about $300 \,\mu g \, m^{-3}$ in the city of

Sheffield. These values are about four times higher than the current (1992) EC recommended limits. By 1983 there had been a 90% reduction in the smoke and sulphur dioxide concentrations. Deaths in the city from bronchitis had fallen by about 70%, matching similar reductions in the rest of the UK. Whilst improved medical treatment and changes in smoking habits have played a part in this reduction in respiratory illness, air-quality improvement has also been a major causal factor. The reduction in smoke and sulphur dioxide concentrations was brought about by both fuel substitution (gas and electricity for coal) and changes in technology (larger electricity generation units and taller flue stacks).

The increased acidity of water, leading to increased levels of dissolved metals, is causing concern in those communities which depend on well water. In large-scale public-supply systems, slaked lime $(Ca(OH)_2)$ is added to the water to raise the pH above 7. This prevents the solution effects on lead and copper pipes that may be found in soft-water areas even without the input of acid rain. Any previously dissolved metals will be precipitated from solution by this treatment. It is more difficult to control the pH in isolated, intermittent water supplies from wells.

Vegetation

Sulphur is an essential nutrient and up to $70 \, kg \, ha^{-1}$ may be needed for optimum growth with intensive agriculture. Forest trees require much smaller quantities of sulphur, being mainly composed of cellulose, and there are few areas where sulphur has been reported as a limiting factor for tree growth. In the UK soils receive 15 to $100 \, kg \, ha^{-1} \, a^{-1}$ from rain and this is usually sufficient for most crops. In others parts of the world (e.g. Australia, New Zealand, and East and Central Africa) sulphur deficiencies occur, and these have become more apparent as sulphur-containing fertilizers – such as ammonium sulphate $((NH_4)_2SO_4)$ and superphosphate $(Ca(H_2PO_4)_2 + CaSO_4)$ – have been replaced by more concentrated, sulphur-free sources of nitrogen and phosphorus.

Direct damage by acid rain or atmospheric sulphur dioxide is largely associated with foliar injury affecting **transpiration**, photosynthesis, leaf protective coatings and leaching of ions from leaves. In the case of agricultural crops there is no clear pattern of decreased yields at the levels of SO_2 found in rural areas. Some studies have shown reductions in yields with possibly 10–20% of UK Grade 1 agricultural land near urban areas affected to give an estimated loss of £220 million per annum (1992 prices). Other studies have shown increases in yields for some crops.

Forests are much more sensitive to damage, mainly because they grow on poorer soils and are in a precarious equilibrium with their environment. As well as direct damage to leaves, the increased rates of removal of essential metals such as magnesium and potassium from the soils and the mobilization of potentially toxic metals such as aluminium, lead, zinc and copper, all produce stress, possibly leading to death (Chapter 10).

Fish

Eggs and fry are especially sensitive and population depletion is most often caused by failure of these juvenile stages to develop rather than by actual fish kills. Incidences of fish kills are usually associated with dramatic sudden increases in acid input, in many cases due to the melting of snow. During the winter the acidic components are trapped in the snow. When the temperature rises in the spring, the contaminated snow melts first and releases the acids in a short concentrated burst that can lower the pH by 1 to 1.5 units to a value of about 4.5. This melting effect is an example of the **depression of freezing point** of a solvent due to the presence of dissolved species. With the reduction in pH, fish suffer from excessive loss of sodium from the gills. The levels of sodium in the blood may be reduced by a half in 2 days, thus upsetting the equilibria in the fish's body fluids. The presence of calcium in the water inhibits this loss of sodium. Different fish species show different sensitivities, with minnows being especially sensitive and eels and pike most resistant to increased acidity. Tolerant species and individuals have been identified, but whether they would grow fast enough to be suitable for restocking commercial fisheries is not known.

As in soils, the presence of acid water produces increased levels of potentially toxic metals in rivers and lakes. Metals at these elevated levels may be taken up by fish and in some cases may play a part in reducing the fish population.

Other aquatic flora and fauna

Molluscs and crustacea rarely survive below pH 6, being more sensitive than fish. In general, species diversity is very much reduced with increased acidity, though some species survive even though fish are absent. Organic decay is much slower in acid lakes and benthic *Sphagnum* species (moss) may cover lake bottoms when calcium is low and light transmission is high.

Buildings and metals

Building stone containing calcium or magnesium carbonates is readily attacked by acid rain with the formation of a soluble sulphate (Eqn 6.15).

$$CaCO_{3(s)} + H_2SO_{4(aq)} \longrightarrow Ca^{2+}_{(aq)} + SO^{2-}_{4(aq)} + H_2O_{(l)} + CO_{2(g)} \qquad (6.15)$$

The building stone may be a limestone or marble (both $CaCO_3$) or it may be a sandstone in which the quartz grains are held together by a coating of calcium carbonate or iron oxide. The iron oxides will also dissolve in acid water (Eqn 6.16).

$$Fe_2O_{3(s)} + 3H_2SO_{4(aq)} \longrightarrow 2Fe^{3+}_{(aq)} + 3SO^{2-}_{4(aq)} + 3H_2O_{(l)} \qquad (6.16)$$

Sulphur dioxide reacts more slowly under dry conditions but overall the effects are the same.

The presence of atmospheric sulphur dioxide, nitrogen oxides and acid rain have all increased the rates of corrosion of metallic structures, such as bridges, and motor vehicles. It is difficult to quantify these effects as there are also other factors such as sea spray and road de-icers which are probably more important causes of accelerated corrosion.

Remedial measures for the sulphur dioxide problem

The major source of the sulphur dioxide responsible for the increase in acid rainfall is the combustion of fossil fuels. The sulphur contents of various fuels are: coal (0.1–6%), fuel oil (0.75–3%), petroleum (0.04%) and diesel oil (0.22%), with natural gas being essentially sulphur free as delivered to the consumer. About 61% of the SO_2 comes from the burning of coal, 25% from oil products, 10% from the smelting of copper sulphide ores and 1.5% from lead and zinc sulphide smelting. The control of sulphur emissions can be achieved by a number of means.

Flue-gas desulphurization

The sulphur compounds are removed from the emitted process gases. This should be easier for the smelting industries where the concentration of sulphur oxides in the gases can be relatively high (5–15%) and direct conversion to sulphuric acid is possible. If the sulphuric acid is not saleable, conversion to elemental sulphur reduces storage problems and allows easier transport to other markets. In fossil-fuel combustion, the concentration of sulphur compounds is usually less than 0.3%. These dilute gases can be scrubbed with alkaline solutions or suspensions of various compounds (Eqns 6.17–6.21).

$$CaCO_3 + SO_2 \xrightarrow{H_2O} CaSO_3 + CO_2 \tag{6.17}$$

$$CaO + SO_2 \xrightarrow{H_2O} CaSO_3 \tag{6.18}$$

$$2CaSO_3 + O_2 + 4H_2O \longrightarrow 2CaSO_4 \cdot 2H_2O \tag{6.19}$$

impure sludge dumped

$$MgO + SO_2 \longrightarrow MgSO_3 \xrightarrow{heat} \underset{reused}{MgO} + \underset{concentrated}{SO_2} \longrightarrow H_2SO_4 \tag{6.20}$$

$$Na_2SO_3 + H_2O + SO_2 \longrightarrow 2NaHSO_3 \xrightarrow{heat} \underset{reused}{Na_2SO_3 + H_2O} + \underset{concentrated}{SO_2}$$

$$\longrightarrow H_2SO_4 \tag{6.21}$$

The products may then be dumped, sold, or the absorbent regenerated and the sulphur compounds dumped or sold. The market for saleable sulphur compounds, such as sulphuric acid, gypsum ($CaSO_4 \cdot 2H_2O$, for plaster board) and ammonium sulphate, is limited. They are all low-cost items and cannot support the high transport costs incurred in delivering them over large distances. In the UK about 750 000 tonnes of sulphur are imported annually compared to an output from electricity power stations of about $1.3 \times 10^6 \, t \, S \, a^{-1}$. More sulphur could be produced than is used thus reducing the price of the sold sulphur compounds. Desulphurization methods that yield throw-away products are much cheaper to operate but produce large quantities of waste that has to be disposed of. A typical British 2000 MW power station would yield about 1 million tonnes of wet sludge or 210 000 t 98% $H_2SO_4 \, a^{-1}$ or 67 000 t S a^{-1} if the desulphurization efficiency was about 90%. The dewatering of the dumped sludge can lead to water-pollution problems and large areas of land are required for storage, e.g. about 1 km × 100 m if the sludge was 10 m deep, for the 10 year output of a 2000 MW power station.

The Drax power station is the largest coal-fired power station in Europe with a gross capacity of 3960 MW. It is being refitted with a flue-gas desulphurization system that will produce about 1 million tonnes per year of commercial grade gypsum when the scheme is completed in 1996. The limestone/gypsum process that is being installed is based on reactions 6.17–6.19. The gypsum is separated from the sludge for sale or for dumping, depending on demand. The sulphur dioxide emissions from the flue stacks should be reduced by about 90% from 350–400 kt a^{-1} to 35–40 kt a^{-1}, provided that there are no breakdowns in the flue-gas desulphurization system. About 11 million tonnes of coal are burned each year at Drax and this yields 2 million tonnes of ash for disposal. The flue-gas desulphurization system will produce an extra 70 000 t of waste sludge (65% solids) and a million tonnes of treatment water each year that will also have to be disposed of. The process will require about 0.7 Mt a^{-1} of limestone to be mined and then transported 70–100 km to Drax. It is clear from these figures that the local environmental impact of this development will be considerable.

Fuel desulphurization

About half the sulphur in coal is present as pyrites, FeS_2, and approximately 80% of this could be removed by grinding up the coal and using a variety of separation techniques. The removal of the organically bound sulphur is much more difficult and requires the conversion of the coal into new liquid or gaseous fuels. Oil can be desulphurized at the oil refinery by the extension of techniques that are used at present for the partial removal of sulphur. In each case the cost of the desulphurization process is high.

Fuel substitution

This includes the use of lower-sulphide fuels such as natural gas, or the substitution of a 1% S coal for a 3% S coal, the replacing of fossil-fuel-fired power stations by nuclear-power stations and the introduction of so-called 'alternative energy' sources, e.g. solar energy, wind, waves, hydropower, geothermal energy (though the last is often associated with sulphurous emissions such as H_2S).

Treatment of affected areas

The liming of land and the addition of limestone to rivers and lakes has been tried. Because of the low economic value of the crops, the large areas to be covered and the need for constant renewal of the applications, these methods do not appear to be feasible except for localized areas of possibly high-yielding agricultural land.

As all remedial methods involve increased costs to producers of sulphur dioxide, there is great reluctance by them to reduce the emissions effectively.

7 Phosphorus

Phosphorus Abundance by weight (the relative abundance is given in parentheses): Earth, 0.2% (10); crust, 1050 ppm (11); ocean, 88 ppb (19).

The solubility and volatility of naturally occurring phosphorus compounds is low, so the geochemical fluxes of the element are mainly dependent upon suspended-solid transfer by rivers to the ocean and upon dust transfer in the atmosphere. Estimates of the various fluxes of phosphorus (Figure 7.1) reflect this low mobility. Many of the values are not very well known because of the heterogeneous nature of particulate matter and the difficulty of adequately sampling the flows. In almost all cases the phosphorus is present as the ortho-phosphate group, PO_4^{3-}, bound to a cation in insoluble inorganic

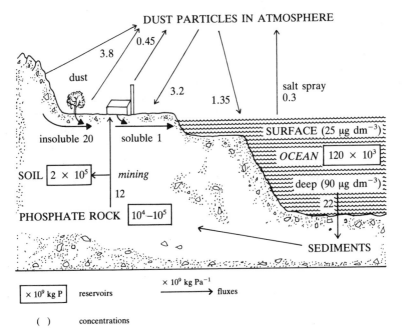

Figure 7.1 The phosphorus cycle.

compounds or as a component of organic molecules. Typical inorganic compounds are the apatites $(Ca_{10}(PO_4)_6(F, OH)_2)$, calcium phosphate $(Ca_3(PO_4)_2)$, aluminium phosphate $(AlPO_4)$ and iron(III) phosphate $(FePO_4)$. Phosphorus is an essential constituent of the energy transferring molecules ATP, ADP and AMP, and of the genetic and information-carrying molecules DNA and RNA. This means that phosphorus is an essential element for all living organisms and it is often a limiting factor in the fertility of soils and aquatic ecosystems. The low solubility of the inorganic compounds limits its availability as a nutrient.

The actual composition of the phosphate minerals is more complex than the formulae given, with a high degree of isomorphous substitution between the major cations Ca^{2+}, Al^{3+}, Fe^{2+} and Fe^{3+}, and the inclusion of relatively large quantities of trace elements (Table 7.1).

The majority of phosphorus added as a fertilizer is rapidly immobilized by inorganic compound formation, and it has been suggested that soluble phosphorus rarely migrates more than 2 or 3 centimetres from a fertilizer granule.

The concentration of phosphorus in lake and river waters is low (60 ppb) and most of the phosphorus transported to the oceans is in particulates and suspended sediments. Of the $20 \times 10^9 \, kg \, P \, a^{-1}$, about $9 \times 10^9 \, kg \, P \, a^{-1}$ are believed to be due to increased erosion caused by deforestation and extensive agricultural activities. Unlike nitrogen, increased concentrations of dissolved phosphorus in water bodies is not due to excess fertilizer in land runoff. The major causes of increased dissolved-phosphorus loadings are sewage disposal and the soluble polyphosphates used in detergents.

Though the phosphate group (PO_4^{3-}) and its related anionic forms $(HPO_4^{2-}$ and $H_2PO_4^-)$ generally have low solubilities, polyphosphates such as the triphosphate ion $(P_3O_{10}^{5-})$ have greater solubilities over a wide range of conditions. The polyphosphates are mainly added to organic detergents to bind the Ca^{2+} and Mg^{2+} ions as soluble polyphosphate complexes. This action prevents the formation of insoluble carbonates with rise in temperature or pH of the water. The latter is important because organic detergents work more efficiently under alkaline conditions. The polyphosphates also act as

Table 7.1 Trace-element concentrations in phosphate rock compared with their concentrations in normal soil (both in $mg \, kg^{-1}$)

	As	Cd	Cr	Hg	Pb	U
phosphate rock	190	100	1600	1000	100	1300
soil	20	0.4	50	0.25	25	1

buffers, helping to stabilize the pH by reacting with acidic components encountered during the washing process. A number of different substitutes for polyphosphates have been suggested, but all have proved to have drawbacks. The substitutes may be less suitable in detergents, e.g. sodium carbonate is too alkaline, or more undesirable environmentally, e.g. sodium nitrilotriacetate (NTA), $N(CH_3COONa)_3$. NTA forms soluble complexes with cadmium and mercury, and produces nitrate on aerobic breakdown; bacterial action on NTA and its breakdown products can form carcinogenic nitrosamine compounds.

Only 20–30% of the soluble phosphates are removed in secondary sewage treatment (Chapter 15), but 90% can be removed during tertiary treatment when the addition of lime together with iron and aluminium compounds causes precipitation of the respective metal phosphates. The residual soluble phosphates in water bodies encourage growth of photosynthetic algae, especially in lakes. The phosphorus concentration in the water decreases as phosphorus is accumulated by the biota and subsequently enters the sediment as the biota die and sink to the bottom of the water body. Either the phosphates are remobilized into the water column as the organic compounds decay, or they are immobilized by inorganic compound formation with Al^{3+}, Ca^{2+}, Fe^{3+} and Fe^{2+}.

The photosynthetic algal population in water bodies is limited by the need for adequate nutrients and light. In lakes, nitrate and phosphate levels are often low and one or other of these may limit the growth of the algae and of the food chain that depends upon these primary producers. If we consider nitrogen and phosphorus only, one phosphorus atom is needed by algae for every 12–20 nitrogen atoms to form the various nitrogen- and phosphorus-containing molecules in their cells. If the N:P ratio in water was 30:1, say, all the phosphorus would be consumed before all the nitrogen (Figure 7.2a), whereas if the N:P ratio was only 6:1, the more rapid removal of the nitrogen would limit biotic growth (Figure 7.2b). However, many studies have shown that if there is adequate phosphorus, nitrogen-fixing blue-green algae develop to increase the N:P ratio effectively and allow further growth. The addition of soluble phosphate allows a large mass of algae to develop, cutting down the light transmission in the water. When the algae die, the dissolved oxygen is removed as the organisms decay. Because the rate of multiplication of the algae is very rapid the term **algal bloom** is applied and there is a rapid change in appearance of the water to a cloudy, greenish soup.

Lakes are especially susceptible to algal blooms as the build-up of nutrients occurs more readily than in the flowing-water conditions of streams and rivers.

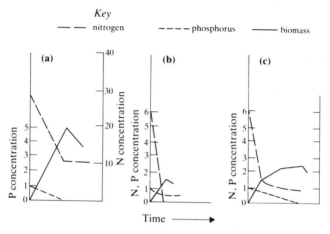

Figure 7.2 The relationship between the development of biomass and the relative concentrations of nitrogen and phosphorus.

Eutrophication is the term given to the enrichment of a body of water with nutrients. There has been an increase in recent years in the rates of eutrophication of lakes and rivers due to the release of (a) nitrates from excess fertilizers and sewage effluent, and (b) phosphates from sewage effluent. Algal blooms in estuaries and coastal waters

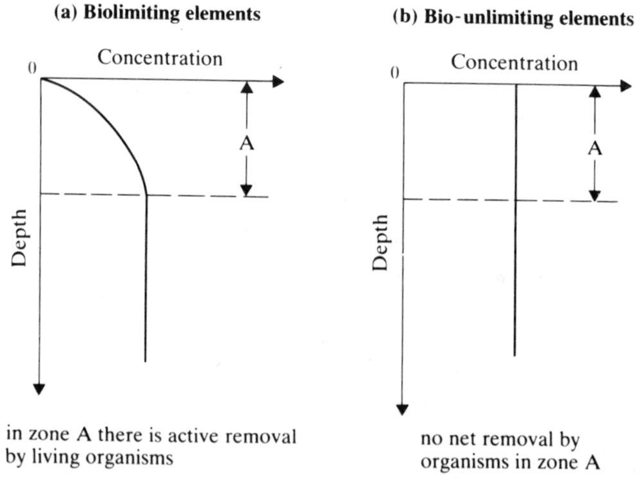

(a) Biolimiting elements

Concentration

0

A

Depth

in zone A there is active removal by living organisms

(b) Bio-unlimiting elements

Concentration

0

A

Depth

no net removal by organisms in zone A

Figure 7.3 The change in concentration with depth in the ocean for (a) biolimiting elements, and (b) bio-unlimiting elements.

have also become increasingly common for the same reasons. In sea water nitrate rather than phosphate is usually the limiting nutrient, the reverse of the freshwater situation.

Nitrogen and phosphorus are described as being biolimiting elements, which means that the concentration of these elements limits biological growth. The variation in concentration of the biolimiting elements with depth in the ocean is characteristic (Figure 7.3a). At the surface, where photosynthesis is at a maximum, the concentration of the biolimiting elements (nitrogen, phosphorus and silicon) is very low. With increasing depth, photosynthesis and elemental uptake decreases and the concentration of the elements in the sea water increases. The bio-unlimited elements, such as sodium, potassium, magnesium, sulphur and chlorine, show no measurable depletion in surface waters (Figure 7.3b).

Part Three

Major elements in the Earth's crust

The crust is the outermost solid layer of the Earth and contains about 0.4% of the mass. It is much richer in silicon and oxygen (Table P1.1) than the Earth as a whole. The crust is very heterogeneous, being made up of a large number of very different rock types. A rock type is defined by the minerals it contains and the supposed conditions under which it formed. A **mineral** is a naturally formed inorganic compound whose composition may be quite complex. Many minerals have a fixed **crystal lattice**, but the atoms, ions or molecular species occupying the lattice positions may differ by greater or lesser amounts from one crystal to the next.

The conditions of formation of rocks can be inferred from the following features.

(a) The minerals present. Each mineral is formed under a characteristic range of temperature and pressure.

(b) The texture of the rock. This is determined by the sizes, shapes and inter-relationships of the minerals in the rock.

(c) The large-scale morphological features. For example, one rock may intrude into another, indicating that it was once molten; or a rock may show ripple marks, indicating that it was deposited as a sediment under water.

The many different rock types are grouped together under three headings depending on how they were formed.

(a) **Igneous rocks** have been formed by the cooling down of a molten mass.

(b) **Sedimentary rocks** have been formed by the accumulation and compaction of minerals and rock fragments at temperatures close to those normal for the Earth's surface.

(c) **Metamorphic rocks** have been formed by the action of increased pressure or temperature or both on igneous or sedimentary rocks.

It has been estimated that the crust contains about 65% igneous rocks, 8% sedimentary rocks and 27% metamorphic rocks. Most of the sedimentary and metamorphic rocks are found in the continental crust and, in fact, about 80% of the surface is composed of sedimentary rocks.

Igneous rocks are thought to form from molten materials, called **magma**, derived from either the upper mantle at mid-ocean ridges, and rifts, where crustal plates are moving apart, or by the remelting of mainly crystalline rocks in subduction zones where plates are converging (Figure P3.1). As the molten material cools, those minerals that are stable at high temperatures will crystallize out. This changes the composition of the remaining magma. Further cooling produces new minerals until all the magma has solidified (Figure P3.2). Which specific minerals form is determined by the composition of the magma at the site of crystallization. Silicon and oxygen being the two most common elements, the major minerals are silicates containing $[SiO_4]$ units linked together by the most common cations – aluminium, calcium, iron, magnesium, potassium and sodium. The higher-temperature minerals are enriched in iron, magnesium and calcium,

(a) Two tectonic plates moving apart – leading to upflow of molten rock from mantle

(b) Two tectonic plates converging – one passes under the other, forming a subduction zone where crustal rocks are remelted

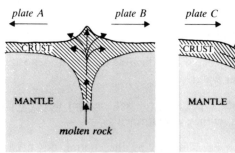
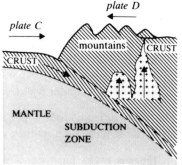

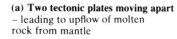 magma formed by remelting of crustal rocks

Figure P3.1 The introduction of magma into crustal regions as a result of plate movements.

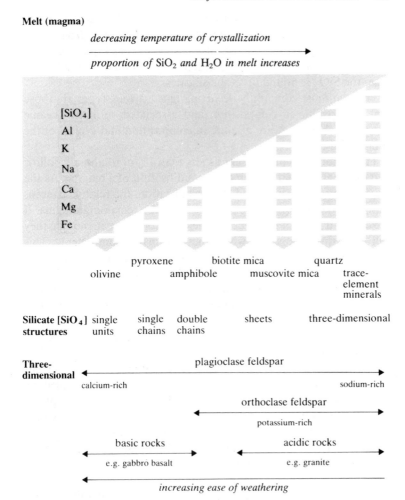

Figure P3.2   The crystallization sequence as magma cools to form a series of igneous rocks.

whereas the lower-temperature minerals contain more sodium and potassium.

Silicates can be thought of as salts of an acidic oxide, SiO_2 (which forms silicic acid, H_4SiO_4), and of basic oxides (e.g. Na_2O, K_2O, CaO and MgO; Eqn P3.1).

$$2MgO + SiO_2 \longrightarrow Mg_2SiO_4 \qquad (P3.1)$$

The apparent proportion of SiO_2 in the rock is used to subdivide the igneous rocks into acidic ($SiO_2 > 66\%$), intermediate ($53–66\%$ SiO_2), basic (45–52% SiO_2) and ultrabasic ($<45\%$ SiO_2) rocks. These subdivisions have nothing to do with the liberation of hydrogen ions and do not imply a specific pH for a particular rock type. In general only acidic rocks actually contain free SiO_2, as quartz. In all other cases the SiO_2 is bound up in the silicates.

The two most commonly occurring igneous rock types are those that are basaltic in composition and associated with the oceanic crust, and those that are granitic in composition and typical of the continental crust.

The minor elements, present as less than 1% of total mass, follow two different pathways during the solidification of a magma. As the major elements are removed by crystallization, the relative concentration of the minor elements increases in the remaining magma. If the concentration of the minor elements becomes great enough, they will form the minerals of which they are major components. This will be particularly important for elements forming minerals with low solubilities. The last remnants of the magma are especially rich sources of minerals of the less common elements, and their crystallization often provides deposits of minerals concentrated enough to be worth mining.

The second pathway followed by minor elements is to become incorporated in the crystal structures of the major minerals. This process is called **isomorphous** substitution and involves one element occupying a position in the crystal lattice normally occupied by another element. The olivine group of minerals shows isomorphous substitution between the major cations Fe^{2+} and Mg^{2+} to give a series of compounds ranging from fayalite ($FeSiO_4$) to forsterite ($MgSiO_4$). The formula is often written '$(FeMg)SiO_4$' to indicate the possible substitution. In addition, replacement of Mg^{2+} (ionic radius 72 pm) and Fe^{2+} (ionic radius 75 pm) by ions such as Ni^{2+} (68 pm), Cu^{2+} (73 pm) and Cr^{3+} (73 pm) may occur so that olivines can contain parts per million (mg kg^{-1}) or parts per thousand (g kg^{-1}) of these elements.

The crystal lattice will remain stable if ions of similar size substitute for each other. The greater the similarity in size, the greater the possibility of substitution. The crystal contains atoms held together either by the transfer of electrons to form ions or by the sharing of electrons to form covalent bonds and molecular species. The substitution of one element for another will occur most readily if this causes minimum disruption to the electron distribution. Fe^{2+} and Mg^{2+} have similar sizes and the same charge, so they often interchange; but Fe^{3+} (ionic radius 55 pm), because of its different charge and radius, does not normally substitute for Mg^{2+}. When there is a difference in oxidation state between substituting atoms, e.g. Ca^{2+} (100 pm) and Na^+ (102 pm) in plagioclase feldspars, the electrical neutrality of the crystal is maintained (a) by substitution of a second element of different and compensating oxidation state somewhere else in the mineral, e.g. $NaAlSi_3O_8$ becomes $CaAl_2Si_2O_8$, where Al^{3+} (39 pm) replaces Si^{4+} (26 pm), (b) by addition of extra ions into a suitably sized hole in the structure, (c) by removal of a balancing species to leave a hole in the structure. Though the general principles guiding substitution in minerals are known, the detailed factors at work in a complex magma or solution are still being disentangled and vary from system to system. The results of these substitution possibilities are that most minerals contain trace amounts (less than 0.01%) of elements that are not normally given in the formula for the mineral. These small amounts can be important sources of elements whose properties may lead to either enhanced or depressed growth of organisms.

Igneous rocks are unstable under the low pressure and temperature conditions of the upper surface of the crust and they break down in the process called weathering (Chapter 8). The fragments are often deposited under the sea in flat beds that on burial become compressed and cemented together to form detrital sedimentary rocks. These rocks are classified on the basis of particle size and this also correlates to some extent with mineral composition. There are also chemical sediments that are formed by precipitation from solution, usually in the ocean. The limestones ($CaCO_3$) and dolomites ($CaMg(CO_3)_2$) are the most abundant examples, with evaporites (largely NaCl, $CaSO_4$ and $CaSO_4.2H_2O$) formed by the evaporation of sea water being the next most common examples.

Contact metamorphic rocks are formed by the alteration of rocks near to igneous intrusions by the action of heat and some pressure. Regional metamorphic rocks are produced by heat and pressure usually due to deep burial or deformation during mountain uplift periods when great stress is applied to the rocks causing new minerals to be formed. If the effects become extreme, the rocks become molten and the difference between metamorphic and igneous rocks is lost.

Though the majority of the crust is composed of silicate minerals, it is often relatively rare minerals that have been concentrated by one of the rock-forming processes that are exploited as sources of raw materials. Examples include sulphide minerals of lead and zinc from the latter stages of magmatic crystallization; bauxite deposits of aluminium from weathering of igneous rocks; lithium phosphate in evaporite deposits. The uncontrolled exploitation of these limited resources will lead to world shortages unless we stop redispersing over the face of the Earth these elements that natural processes have taken so long to concentrate. The second law of thermodynamics indicates that disorder will arise from order but there seems to be little to gain from hastening the change!

8 Silicon

Silicon Abundance by weight (the relative abundance is given in parentheses): Earth, 14.3% (3); crust, 29.5% (2); ocean, 2.9 ppm (14).

8.1 Silicate minerals

The majority of the Earth's crust is composed of silicate minerals, and these minerals have been classified on the basis of how the $[SiO_4]$ units are arranged (Figure P3.2). The individual $[SiO_4]$ unit (Figure 8.1) can be considered to consist of four closely packed oxide ions, O^{2-}, with a silicon ion, Si^{4+}, occupying the tetrahedral hole between them. In practice the Si–O bonds have a large degree of covalent character. The Si^{4+}–O^{2-} bond has been estimated to be 52% covalent. The Si–O bond (bond energy: $372\,kJ\,mol^{-1}$) is over twice as strong as the Si–Si bond (bond energy: $180\,kJ\,mol^{-1}$); therefore the Si–O arrangement is the preferred structure wherever, as in the Earth's crust, there is sufficient oxygen present to allow its formation. Each oxygen can form bonds with two atoms of silicon, and so adjacent $[SiO_4]$ units can be joined together by the sharing of oxygen atoms to give chains, sheets or three-dimensional networks.

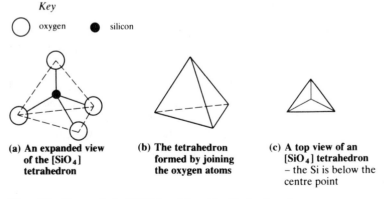

Key

◯ oxygen   ● silicon

(a) An expanded view of the $[SiO_4]$ tetrahedron

(b) The tetrahedron formed by joining the oxygen atoms

(c) A top view of an $[SiO_4]$ tetrahedron – the Si is below the centre point

Figure 8.1 The tetrahedral $[SiO_4]$ unit found in silicate minerals.

The various minerals are characterized by the structural arrangement of $[SiO_4]$ units and the major cations present, but in most silicate minerals there tends to be widespread substitution of one element by another of similar size. If all four oxygens of one $[SiO_4]$ unit are shared with other $[SiO_4]$ units, the overall formula becomes SiO_2 and the product, quartz, is a giant molecule. When aluminium, Al^{3+}, replaces some of the silicon, Si^{4+}, the structure requires further cations – such as calcium, Ca^{2+}, potassium, K^+, or sodium, Na^+ – to regain electrical neutrality. Such minerals are called aluminosilicates. Examples are the plagioclase feldspar group, $(Ca, Na) (Al, Si) AlSi_2O_8$, and potash or orthoclase feldspar, $KAlSi_3O_8$.

The structural features of the silicate minerals (Figure P3.2) are often reflected in their bulk properties. Mica, for instance, separates into thin sheets due to cleavage parallel to the structural layers which are only weakly held together; and blue asbestos, one of the amphiboles, is fibrous due to cleavage parallel to the chains of silicate units.

8.2 Weathering

The solid rocks that compose the bulk of the crust do not remain compact, coherent masses when exposed to the atmosphere. In many areas they are covered by a soft particulate layer called **soil**. Even where the solid rocks appear uncovered by soil or other fragmented material, close examination reveals that the surface layers have a slightly different appearance and composition from those of inner regions. Also, over a period of time, the outer layers are being removed. This disintegration of solid rocks is part of the weathering process in which the rocks and minerals are altered until they are chemically and physically in equilibrium with their surroundings.

It is convenient when looking at the mechanism of weathering to divide the process into two mechanisms that operate together and reinforce each other's actions.

(a) Physical, or mechanical, weathering, in which the rock and subsequent fragments are broken into smaller particles and transported from one locality to another, lower site.
(b) Chemical weathering, in which the chemial composition of the rocks and fragments is changed to produce more stable compounds or species.

The mechanical breakdown into smaller particles produces a larger surface area over which chemical reactions can occur and the changes in chemical composition tend to weaken the structures so that they are easier to break up.

For the environmental chemist, one of the most important features of weathering is that it mobilizes many chemical substances. It is this mobility aspect that we shall concentrate on. In general the lower the solubility of a compound the smaller will be the interaction between that compound and a live organism. Therefore, processes like weathering that affect the solubility of elements are of prime importance in controlling the viability of organisms. There is a wide variety of often quite complex reactions occurring during weathering and in this book we shall only illustrate some of the general principles by considering aspects of the weathering of igneous rocks.

The composition of the parent rock and the weathering conditions together control the proportion of soluble and insoluble products formed during weathering. The majority of the minerals in igneous rocks were formed at high temperatures and usually high pressures. In general the higher the apparent temperature of formation of the mineral, the more rapidly it is likely to weather (Figure 8.2). The more ionic a bond is, the more soluble in water the mineral is likely to be. The Si–O bond has more covalent character than the metal–oxygen bonds in minerals. It is found that minerals in isostructural groups with the greatest proportions of $[SiO_4]$ units are the most stable, e.g. sodium-rich plagioclases have 61–69% SiO_2 and are much more stable than the calcium-rich plagioclases with 43–51% SiO_2.

The isomorphous substitution of one element for another in silicate lattices, which is widespread, can cause either an increase or a

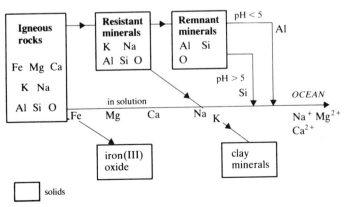

Figure 8.2 The general processes that occur during the weathering of igneous rocks.

decrease in stability of the crystal structure. The consequence is that the individual crystals of a mineral in a rock may weather at very different rates, depending upon the substituted elements. Also the trace elements released by this weathering can vary markedly from rock to rock, even though the mineral composition appears very similar.

Figures 8.2 and 8.3 illustrate the general trends during weathering, and some of the breakdown products for individual minerals. It should be noted in particular that water is a prerequisite for these reactions: in many arid areas the amount of water is so low that physical weathering is the dominant mechanism. The reactivity of aqueous solutions is increased by the presence of hydrogen ions, which aid the dissolution of metallic elements. The acidity of the water can arise from three major sources.

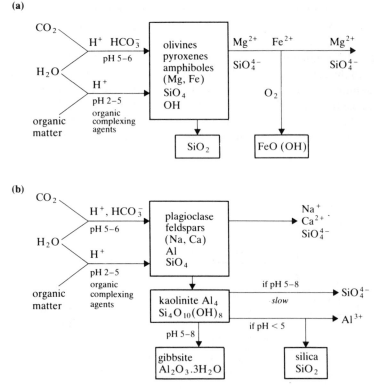

Figure 8.3 The chemical weathering of a number of common minerals found initially in igneous rocks.

(a) Gases dissolved from the atmosphere. The main example is carbon dioxide, which gives an equilibrium pH of 5.7, but in some areas of the world (e.g. Northern Europe), the contribution from sulphur dioxide has become even more important and sometimes the pH has dropped to about 4.

(b) The oxidation of plant remains (Chapter 4). This can increase the concentration of carbon dioxide in the soil atmosphere (the gas occupying the space between the soil grains) by a factor of a hundred.

(c) The formation of organic acids during the decay and partial oxidation of plant remains (Eqn 8.1). This can lead to a pH in the range of 4 to 5.

$$\text{organic C} + O_2 \longrightarrow \underset{\substack{\text{carboxylic}\\\text{acid}}}{RCOOH} + \underset{\text{phenols}}{ArOH} \longrightarrow RCOO^- + H^+$$
$$+ ArO^- + H^+ \qquad (8.1)$$

Reaction of the acidic solution with the silicate minerals reduces the acidity of the aqueous solution (Eqn 8.2), and it may even become neutral or alkaline.

$$\underset{\substack{\text{acidic}\\\text{solution}}}{2H^+_{(aq)}} + M_2SiO_4 \longrightarrow \underset{\substack{\text{altered}\\\text{silicate}}}{H_2MSiO_4} + \underset{\substack{\text{neutral}\\\text{solution}}}{M^{2+}_{(aq)}} \qquad (8.2)$$

These changes in pH may also change the solubility of the species present (Figure 8.3). If the pH is between 5 and 9, silicon is more soluble than aluminium and iron, and will be leached out of a rock or soil to leave behind a mixture of hydrated aluminium and iron oxides which will form a lateritic soil (more aluminium oxide) or feralitic soil (more iron oxide). When the pH is below 5, aluminium becomes more soluble than silicon and will be leached out. The solubility of iron also increases at low pH, especially if the ground becomes waterlogged for part of the year and reducing conditions are produced (Chapter 9). Low pH, as it is usually associated with organic matter, tends to occur close to the surface of a soil or rock. As the acidic water percolates through the soil, reacting with the minerals, the pH rises and the solubility of iron and aluminium is reduced. The aluminium and iron compounds precipitate out of solution and often form a distinctive layer, called the B horizon.

The part played by the decay of organic matter in the formation of carbon dioxide and organic acids in soils has been described. As well as being responsible for lowering the pH and making the soils more acid, the organic decomposition processes can affect the progress of weathering in other ways. One major effect can be to change the redox conditions. Any such changes bring about changes in the solubility of elements such as iron (Chapter 9), sulphur (Chapter 6) and nitrogen

(Chapter 5). Many of the organic compounds produced during decomposition can act as complexing agents which may increase or decrease the solubility of an element. Organic compounds with high relative molecular masses, such as those that make up peat and the humose layers in soils, are insoluble in water. Therefore metals, e.g. lead, forming **complexes** with these substances are held in immobile forms in surface layers of soils (Chapter 13). Compounds with lower relative molecular masses (values less than 1000), and which contain OH or COOH groups, are more likely to form soluble complexes with metals and so increase the mobility of the metal. The binding of the organic compound to the metal ion in the complex prevents the metal ion reacting with other chemicals. Thus the chemistry of the complexed ion can be completely different from the chemistry of the simple ion. For instance, iron(III), Fe^{3+}, will often precipitate from solution (Eqn 8.3) when the pH rises above 4 (Chapter 9). If oxalic acid, HOOCCOOH, is present an iron–oxalic-acid complex such as trioxalatoiron(III)ate, $[Fe(OOCCOO)_3]^{3-}$, is formed that is soluble at pHs higher than 4 (Eqn 8.4).

$$Fe^{3+} + 3H_2O \longrightarrow \underset{\text{insoluble}}{Fe(OH)_3} + 3H^+ \tag{8.3}$$

$$Fe^{3+} + 3HOOCCOOH \longrightarrow \underset{\text{soluble}}{[Fe(OOCCOO)_3]^{3-}} + 6H^+ \tag{8.4}$$

One of the great difficulties in studying natural solutions is knowing in what form an element is actually present. Is the copper in the soil-pore water a simple hydrated ion or is it in a complex with, for example, citric acid, as $[Cu(C_6H_4O_7)]^{2-}$? The interaction of the organic compounds, produced by microbial action, with the minerals in rocks and soils can prove to be the dominant factor in determining the rate at which weathering proceeds. This is because of the way in which chemical properties, and especially solubility, can be altered. The chemical reactions that occur are often very complicated and there is little detailed knowledge of the actual changes occurring in natural systems.

A characteristic feature of soils is the formation of a series of layers of different composition, called **horizons** (Figure 8.4), parallel to the daylight surface. The thickness, composition and appearance of the various horizons varies widely from place to place, being determined by the interaction of source material, climate, topography, biological activity and age. The general pattern found shows movement of material away from the surface. The movement is initially downwards, but may then become lateral with the ground-water flow. Intermediate stages result in the accumulation of particular types of matter in certain horizons (e.g. organic matter in Ao horizons; iron

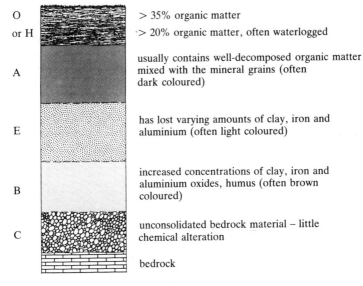

Figure 8.4 Diagrammatic representation of an idealized soil profile.

oxides, clays and co-precipitated trace elements in the B horizon) or in the removal of materials from certain horizons (e.g. silicon or aluminium, depending upon pH, from the E horizon).

The general changes occurring in silicate-based rocks during weathering and soil formation can be summarized as:

(a) breakdown into smaller particles;
(b) formation of new minerals that have a greater proportion of octahedral structures stable at low temperatures and pressures;
(c) the preferential loss of silicon, found in tetrahedral units, compared to aluminium and iron, found in octahedral units;
(d) increasing numbers of OH^- ions, or even H_2O groups, compared to O^{2-} ions in the new minerals;
(e) preferential loss of sodium, calcium and magnesium compared to potassium, aluminium and iron;
(f) general mobilization of trace elements as the original silicate minerals disintegrate (though these elements may be trapped again in the new low-temperature minerals).

Ultimately both the soluble components ($2.5-4 \times 10^9 \, t \, a^{-1}$) and the particulate residues ($6.5-18 \times 10^9 \, t \, a^{-1}$) from weathering will be transferred to the oceans. The most important transfer agents are the rivers (89%), but glaciers (7%) and ground water (2%) both provide significant inputs to the ocean. Dust (0.2–0.5%) blown into the atmosphere can be of considerable local importance. The rivers have been estimated to have an average ratio of dissolved-to-suspended material of 1:2.6, but higher transfer rates and ratios (1:4.5) have also been suggested The actual ratio of dissolved-to-suspended load varies greatly from river to river and even from continent to continent. About 80% of all suspended material entering the world's oceans comes from South-East Asia. The distribution of the dissolved load entering the oceans is much more evenly spread between the conti-nents, but Australia and Antarctica provide very low inputs due to their arid and frozen natures respectively.

9 Iron

Iron Abundance by weight (the relative abundance is given in parentheses): Earth, 34.8% (1); crust, 5.0% (4); ocean, 3.4 ppb (28).

Assuming that the core is mainly composed of liquid iron and that the mantle contains a high proportion of **ferromagnesian silicates**, both will have higher concentrations of iron than that in the crust. In addition, the oxidation state of the iron changes from 0, the free metal, to 2 (called iron(II) or ferrous iron) in ferromagnesian minerals, and to 3 (called iron(III) or ferric iron) in many crustal rocks.

The oxygen-rich atmosphere of the Earth ensures that both iron(0) and iron(II) are unstable under normal conditions at the Earth's surface and both are oxidized to iron(III) (Eqns 9.1 and 9.2) producing the characteristic red-brown yellow colours of the various forms of iron(III) oxide.

$$4Fe + 3O_2 \xrightarrow{\hspace{1cm}} 4Fe^{3+} + 6O^{2-} (\xrightarrow{\hspace{1cm}} Fe_2O_3) \tag{9.1}$$

$$4Fe^{2+} + O_2 \xrightarrow{\hspace{1cm}} 4Fe^{3+} + 2O^{2-} (\xrightarrow{\hspace{1cm}} Fe_2O_3) \tag{9.2}$$

The low solubility of iron(III) oxides means that the cycle of iron (Figure 9.1) is dominated by solid-transfer reactions and most of the iron reaching the oceans is carried by rivers as suspended solids. Of the estimated $1.1 \times 10^{12} \, kg \, a^{-1}$ iron transported by rivers to the oceans, only $1.4 \times 10^9 \, kg \, a^{-1}$ is thought to be in solution.

9.1 Iron in natural systems

The lithosphere

The changes in oxidation state of iron in the Earth's crust are of great importance in understanding its properties, including mobility. Many of the factors that are important for iron are also important for other elements with variable oxidation states – manganese, for instance, is very similar to iron. The relative stabilities of iron(II) and iron(III) are such that only small changes in natural environmental conditions can cause either iron(II) to be oxidized to iron(III) or iron(III) to be reduced to iron(II), with consequent large changes in solubility. Changes

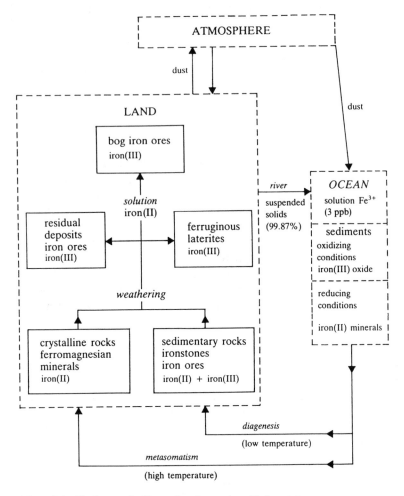

Figure 9.1   The iron cycle, illustrating changes in oxidation state.

in either pH (i.e. proton activity) or redox potential, Eh (which is a measure of how oxidizing or reducing a system is, i.e. it is a measure of electron activity) are effective. The Eh–pH diagram (for example, see Figure 9.3), also known by engineers as a Pourbaix diagram, is a useful way of illustrating which forms of an element may be stable under specific conditions of pH and Eh. The diagram contains so-called regions of dominance, inside whose boundaries one form is the most stable under that set of Eh–pH conditions.

When considering natural systems on the surface of the Earth, only a limited range of pH and Eh need be investigated. Because water is stable on the Earth's surface, redox potentials that would cause either the oxidation of water (Eqn 9.3), i.e. high values of Eh, or reduction of water (Eqn 9.4), i.e. low values of Eh, must not occur. In practice the natural range of Eh values is even narrower, being mainly confined to the region between + 0.6 V and − 0.8 V, with surface water being in

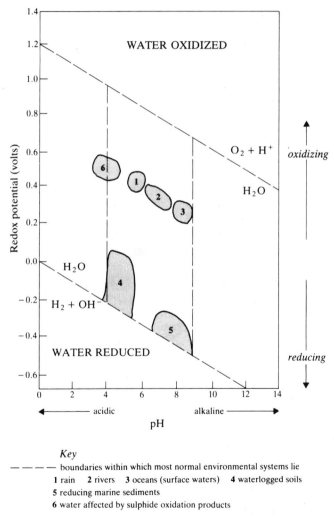

Key

— — — — boundaries within which most normal environmental systems lie

1 rain **2** rivers **3** oceans (surface waters) **4** waterlogged soils

5 reducing marine sediments

6 water affected by sulphide oxidation products

Figure 9.2 The Eh–pH conditions of some natural environments.

the range $+0.2$ V to $+0.5$ V (Figure 9.2).

$$2H_2O \rightleftharpoons O_2 + 4H^+ + 4e^- \tag{9.3}$$

$$2H_2O + 2e^- \rightleftharpoons H_2 + 2OH^- \tag{9.4}$$

The normal range of pH of waters in natural systems is 4 to 9, being controlled by the $CO_2/HCO_3^-/CO_3^{2-}$ system. Occasionally, low values below pH 4 may be found, usually due to the oxidation of sulphide minerals, either as a natural process (often microbially mediated) or as a result of mining which brings sulphide minerals into contact with the atmosphere. Values higher than pH 9 are associated with alkaline environments, mainly due to the presence of sodium carbonate in desert regions.

The Nernst equation (Eqn 9.5) provides a more formal definition of redox potential, Eh:

$$Eh = E^0 + \frac{RT}{nF} \ln\left(\frac{[\text{oxidized species}]}{[\text{reduced species}]}\right) \tag{9.5}$$

where   Eh = equilibrium redox potential, in volts, compared to a standard hydrogen half-cell, which has a value of 0.0 V;
E^0 = standard reduction potential at pH = 0, 25 °C, 1 atm, and unit activities of solute ions;
R = gas constant, 8.314 J mol^{-1} K^{-1};
T = absolute temperature, degrees K;
n = number of electrons transferred;
F = Faraday constant, $96\,487$ C mol^{-1};
[oxidized] [reduced] = activities of oxidized and reduced species, respectively.

Though the Eh value is the value compared to the hydrogen electrode, in the majority of measurements other electrodes are used and a conversion factor applied.

Hydrogen ions are released when water is oxidized (Eqn 9.3), therefore changing the H^+ activity (or H^+ concentration) of H_2O and O_2. This dependence of Eh on pH is shown below when values typical of the conditions on the Earth's surface are substituted in Equation 9.6.

$$Eh \text{ (for Eqn 9.3)} = E^0 + \frac{RT}{4F} \ln\left(\frac{[O_2][H^+]^4}{[H_2O]^2}\right) \tag{9.6}$$

As the water can be considered to be pure, its activity is constant with a value of 1.

At a temperature of 298.15 K (25 °C), $E^0 = +1.23$ V. Thus

$$\text{Eh} = 1.23 + \left(\frac{8.314 \times 298.15}{4 \times 96487} \right) \times 2.303 \times \log([O_2][H^+]^4)$$

$$= 1.23 + 0.0148 \times \log([O_2][H^+]^4)$$

$$= 1.23 + 0.0148 \times \log[O_2] + 4 \times 0.0148 \times \log[H^+]$$

$$= 1.23 + 0.0148 \log[O_2] + 0.059 \log[H^+]$$

But

$$\log[H^+] = -\text{pH}$$

and at the Earth's surface

$$[O_2] = \text{partial pressure of } O_2 \text{ in the Earth's atmosphere}$$

$$= 0.21 \text{ atmospheres}$$

Therefore

$$\text{Eh} = 1.23 + 0.0148 \log(0.21) - 0.059 \text{ pH}$$

$$= 1.23 - 0.01 - 0.059 \text{ pH}$$

$$= 1.22 - 0.059 \text{ pH}$$

For iron the simplest approximation to natural conditions is illustrated by the Fe–H_2O system (Figure 9.3), which indicates the boundaries of the regions of dominance of the various species at two concentrations of iron. At the boundary line between two regions of dominance the activities of the two species are equal, but the activity of a species decreases very rapidly outside its own region of dominance. The region of iron(III) hydroxide stability includes much of the Eh–pH range found in natural systems (Figure 9.2). The stability region for metallic iron lies below the water stability region, indicating that metallic iron is unstable and will tend to convert to Fe^{2+} or $Fe(OH)_3$.

Rain water running through a soil containing organic matter could have a pH of 4.5 and Eh of $+0.5$ V (position A in Figure 9.3), and any iron minerals in the soil will tend to dissolve. As the water flows through the soil, reaction with other minerals, such as carbonates, in the soil will raise the pH to a value of 6, say (position C in Figure 9.3). However, when the pH approaches 5, the iron will precipitate as iron(III) hydroxide (position B). Thus the iron will have been transported from the mineral near the surface of the soil to the region of higher pH that may be a few millimetres, centimetres or metres away. If there was a lot of rain and the soil became waterlogged, the water would occupy all the pores in the soil, displacing the air and its

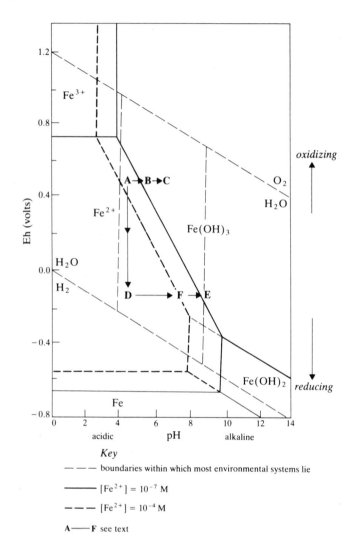

Key

- – – – boundaries within which most environmental systems lie

——— $[Fe^{2+}] = 10^{-7}$ M

– – – – $[Fe^{2+}] = 10^{-4}$ M

A———F see text

Figure 9.3 The Eh–pH diagram for the iron–water system. (After Garrels, R. M. and L. Christ 1965. *Solutions, minerals and equilibria.* New York: Harper & Row.)

associated oxygen. Microbial respiration would use up the small amount of dissolved oxygen in the water and conditions would become anaerobic. The Eh might drop to -0.1 V (position D). As the water moves through the soil, the pH would again rise, but now the $Fe(OH)_3$ boundary would not be reached until pH 9 (position E). In waterlogged reducing conditions, the transport of soluble iron(II) is likely to be more extensive than in aerobic conditions. In practice, because more Fe^{2+} will be dissolved, the iron molarity is likely to increase from about 10^{-7} M $(5.6\,\mu g\,dm^{-3})$ to about 10^{-4} M $(5.6\,mg\,dm^{-3})$ and the $Fe(OH)_3$ boundary will be between pH 7 and 8 (position F). If there was much water movement, this would still lead to extensive leaching. The leaching of iron in waterlogged soils is the

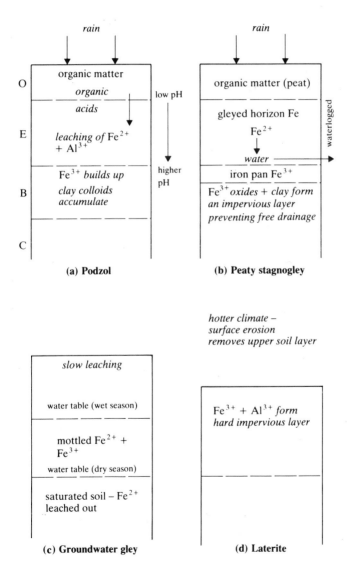

Figure 9.4 The formation of podzolic, gleyed and lateritic soils.

characteristic feature of gleying and in acid soils in the formation of podzols (Figure 9.4).

In natural systems there are many different dissolved species other than iron, and these can influence the behaviour of iron by reacting with it to form either relatively insoluble compounds (e.g. sulphide or phosphate) or stable complexes (e.g. organic acids with iron(III)) that prevent the iron being precipitated. Diagrams including all possibilities would be very complicated, but an approximation to many natural conditions is given by the Eh–pH diagram for the $Fe–CO_2–S–H_2O$ system (Figure 9.5). The normal range of conditions found in soils is outlined and it can be seen that haematite (Fe_2O_3), siderite ($FeCO_3$), pyrites (FeS_2) and soluble Fe^{2+} are all possible phases. In general, the most common iron species are Fe^{2+} in acidic reducing conditions and haematite in less acidic and in alkaline oxidizing conditions. Though haematite is the most stable oxide form, goethite ($FeO \cdot OH$) and other hydrated oxides are almost as stable and just as insoluble. The transformation from one of these hydrated solids, which may be precipitated first, to haematite is very slow. It is often found that haematite is formed in warm wet regions and goethite elsewhere.

Because the boundary between the Fe^{2+} and Fe_2O_3 domains is just on the acid, reducing side of normal surface waters, relatively small changes in pH or Eh can cause either marked dissolution or precipitation of iron. For example, ground waters tend to be reducing and contain several milligrams per cubic decimetre of dissolved Fe^{2+}, but on emerging from a spring the Eh will rise rapidly as oxygen is dissolved (movement from A to B in Figure 9.5) and a red-brown precipitate of iron(III) oxide will form. The first-formed precipitate is usually a gelatinous hydrated iron(III) oxide, $Fe_2O_3 \cdot nH_2O$, that gradually dehydrates on ageing, though formation of Fe_2O_3 may take years. When in the freshly precipitated, highly hydrated form, the iron(III) is more likely to redissolve than when in the partially or fully dehydrated form. The surface area of the fresh precipitate is very high and the stoichiometric charge balance is not very good. As a result many other ions, particularly other metals, may be sorbed and co-precipitated (Figure 9.6). The precipitate acts as a scavenger for these metals and tends to collect in the soil's B horizon, either because the pH is higher in this horizon or because it acts as a barrier to the transport of colloidal iron(III) oxide particles (size less than 1×10^{-6} m) by ground water.

The most noticeable effects of iron redox reactions in soils are seen in podzolization, gleying and laterite formation. A podzol is typically formed in a cool temperate climate in which there is sufficient rainfall

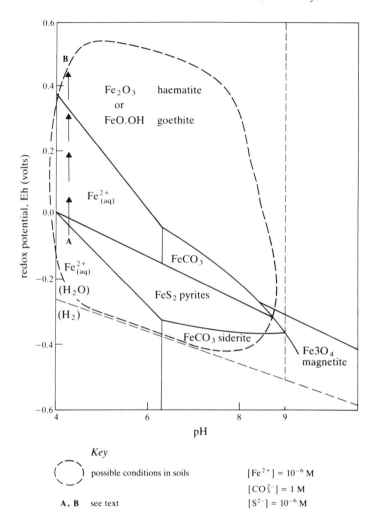

Figure 9.5 The Eh–pH diagram for the Fe–H$_2$O–CO$_2$–S system. (After Garrels, R. M. and L. Christ 1965. *Solutions, minerals and equilibria*. New York: Harper & Row.)

to maintain downward leaching and in which the breakdown of organic matter is slow (Figure 9.4a). The acid conditions favour the dissolution of Fe^{2+} and Al^{3+} and their transport from the A horizon, together with organic matter, to the B horizon. Here the higher pH favours precipitation of iron(III) oxide, together with the accumulation of clay and other colloids. The result is a grey silica-rich A horizon and a brownish B horizon. If the B horizon is relatively

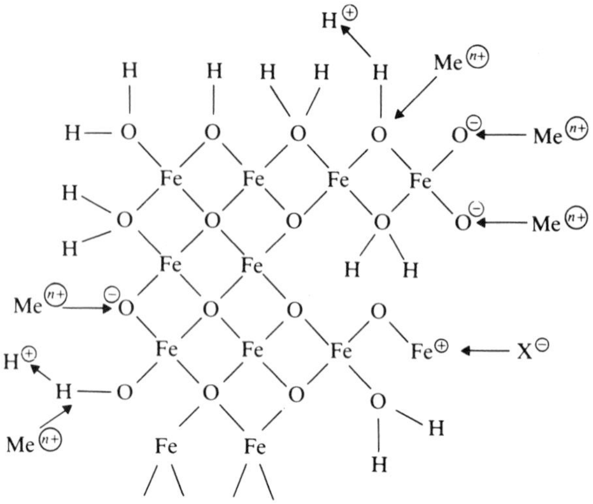

Key

X⁻   anion. e.g. PO_4^{3-}

Me^{n+} cation. e.g. Pb^{2+}. Zn^{2+}

Figure 9.6 An illustration of surface processes associated with hydrated iron(III) oxide.

impervious (maybe due to clay), the upper portion of the soil will tend to become waterlogged (Figure 9.4b) and the reducing conditions that develop (gleying) promote the further leaching of iron. The periodic drying out of the upper layers favours the formation of an iron pan (a hard impervious layer of iron(III) oxide) at the top of the B horizon. In soils with an oscillating water table near the surface, the gleyed regions may be below the surface (Figure 9.4c). This zone is typically mottled due to the presence of Fe^{2+} and Fe^{3+}, whereas the permanently saturated zone underneath contains Fe^{2+} that is being leached away. In hotter climates the dropping of the water table may cause most of the iron in this zone to oxidize rapidly to iron(III) and form an extensive iron pan (Figure 9.4d). This layer also accumulates aluminium oxides and, as it is impervious, subsequent heavy rainfall tends to wash off the looser soil above, leaving the hard lateritic layer exposed.

The biosphere

The concentration of iron in the biosphere is much lower than in the lithosphere, but iron is essential to the functioning of all organisms

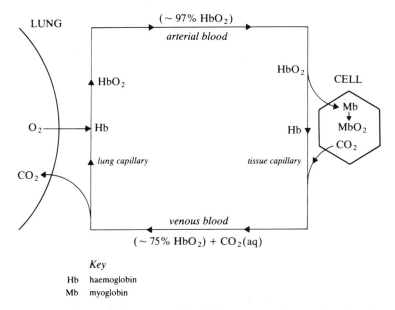

Figure 9.7 Diagrammatic representation of the transport of oxygen from lungs to body cells by blood.

and it has been suggested that iron deficiency is probably the most prevalent deficiency state affecting human populations. As we have seen already, though iron is a common constituent of soils, levels in solution are generally low and it is not very available.

One of the functions of iron is to act as an oxygen carrier in vertebrates. Oxygen is carried to the cells of the body by haemoglobin molecules concentrated in the red blood cells of the bloodstream (Figure 9.7). The presence of haemoglobin (Hb) raises the oxygen-carrying capacity of the blood from about $3\,cm^3\,dm^{-3}$ to $200\,cm^3\,dm^{-3}$. The red blood cells, erythrocytes, are $8\,\mu m$ in diameter and each contains about 280×10^6 molecules of haemoglobin. The cells of the body contain myoglobin, which accepts and stores the oxygen carried by the haemoglobin. Both haemoglobin and myoglobin (Figure 9.8) contain iron(II) bound to the four pyrrole nitrogen atoms of protoporphyrin IX. This arrangement is called the haem group. The iron is also bound to a nitrogen atom from a histidine group that is part of a protein chain and called the globin part of the molecule. Haemoglobin contains four of these units and binds four molecules of O_2, one to each iron, whereas myoglobin consists of only one unit and binds only one O_2 molecule. Myoglobin has a greater affinity for oxygen than has haemoglobin. The affinity of haemoglobin

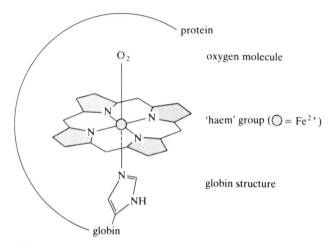

Figure 9.8   Structure of oxygenated myoglobin.

for O_2 varies with pH, being lower at lower pHs (higher carbon dioxide concentrations in the tissues). This helps the transfer of O_2 to myoglobin, and when the CO_2 carried by the blood is released to the lungs, the lower CO_2 and higher O_2 concentrations favour O_2 pick-up. The globin reduces the ability of H_2O and O_2 molecules simultaneously to approach the iron(II), which would then be oxidized to iron(III). The iron(III) complexes, methaemoglobin and metmyoglobin, do not bind O_2, which explains the reduced O_2 transport induced in infants with methaemoglobinaemia (Chapter 5). The globin is not completely successful in preventing methaemoglobin production, but an enzyme reduction mechanism keeps the level to about 1%.

Carbon monoxide, CO, complexes more strongly than does O_2 with the Fe^{2+}, forming carboxyhaemoglobin (HbCO; Eqn 9.7).

$$HbO_2 + CO \rightleftharpoons HbCO + O_2 \qquad (9.7)$$

If someone suffering from carbon monoxide poisoning is given pure oxygen to breathe, the equilibrium reaction 9.7 will be moved to the left and the conversion of HbCO to HbO_2 will be promoted. Whether this treatment proves successful depends upon the extent of oxygen deficiency in the body before treatment commences.

The iron(II)–iron(III) transition (Eqn 9.8) is made use of by many molecules involved in electron-transfer reactions, such as the ones we came across when discussing the role of nitrogenase in the fixation of nitrogen.

$$Fe^{2+} \rightleftharpoons Fe^{3+} + e^- \qquad (9.8)$$

Animals, including humans, have only a limited ability to excrete metabolic iron, so the correct iron balance in the body is mainly achieved by adjustments to the rate of absorption. Normal adults only absorb about 5–10% of the iron in a balanced diet, whereas iron-deficient individuals absorb 15–20%, or more, of the iron. The iron has generally to be in the iron(II) state and it tends to be more readily absorbed if already bound to haem groups. The detailed mechanism of absorption is not understood: there appear to be many interacting and interfering factors. The majority of the iron in the body (about 4 g in an adult) is continuously recycled, thus counter-balancing the relative difficulty of absorption. Iron deficiency arises either from poor diets (intake less than $10–12\,\mathrm{mg\,d^{-1}}$) or by excessive loss, usually due to bleeding. The deficiency reduces the level of haemoglobin in the blood, producing anaemia, which in turn reduces the transport of oxygen for energy production in the cells, leading to listlessness.

9.2 Iron in industrial systems

Iron is the most commonly used metal, with iron production at about 540 million tonnes per annum and steel production 780 million tonnes per annum. (Steel production is higher due to the recycling of scrap steel and to the other metals added to iron to make the steel.)

Iron is produced by the reduction of various ores (haematite, Fe_2O_3; goethite, $FeOOH$; magnetite, Fe_3O_4; siderite, $FeCO_3$), but iron(III) oxide is by far the major component. One of the reasons iron has been used by humans for so long is that this reduction is fairly easily achieved using carbon (Eqn 9.9) and associated compounds.

$$2Fe_2O_3 + 3C \longrightarrow 4Fe + 3CO_2 \tag{9.9}$$

Charcoal was used initially, but nowadays coke is the source of carbon.

The reduction is carried out in a blast furnace which is in essence a large steel cylinder (maybe 30–40 m high, 10–17 m diameter) lined with refractory bricks to withstand the high temperatures. Iron ore, coke and limestone are added from the top, through a double bell which prevents the escape of poisonous gases; and hot air, often partially enriched in oxygen and some hydrocarbon fuel, is blown in from near the bottom.

The initial reactions of coke and hot air (Eqns 9.10 and 9.11) produce enough heat to support the other reactions in the blast furnace, provided that the initial air temperature is above 500 K. (Usually it is 1200–1600 K.)

$$C + O_2 \longrightarrow CO_2 \tag{9.10}$$

$$2C + O_2 \longrightarrow 2CO \tag{9.11}$$

The CO_2 reacts with more coke, producing more carbon monoxide (Eqn 9.12).

$$C + CO_2 \longrightarrow 2CO \tag{9.12}$$

Any CO_2 formed by the reduction of iron(II) oxide (Eqn 9.13) is reconverted to CO above 1200 K (Eqn 9.12).

$$FeO + CO \longrightarrow Fe + CO_2 \tag{9.13}$$

At lower temperatures the proportion of CO_2 in the gas increases until the gas leaves the blast furnace. The presence of unoxidized CO means that the gas has about one-tenth of the energy of an equal volume of natural gas, and this energy is utilized by burning it so that it either pre-heats (via heat exchangers) the air entering the blast furnace, or drives plant and electrical generators on the site.

The iron from the blast furnace is called pig iron, and the majority of it is kept molten and used to make steel. Because of its high carbon content, pig iron is very brittle; though it can be cast it cannot be rolled or worked easily.

The slag is a complex mixture of compounds such as calcium aluminosilicates formed by the reaction of lime, CaO (a basic oxide), with the various acidic oxides of silicon, aluminium, manganese, phosphorus, etc. Oxide impurities are concentrated in the slag but any impurities reduced to the elemental stage (as occurs near the bottom of the furnace) are more soluble in the molten metal. After tapping it from the furnace the slag is dumped, with varying amounts being used for bricks, concrete, ballast and insulation wool.

Steels are alloys of iron containing quantities of other elements that modify the properties of iron. The quantities are controlled to produce the desired material, whether a ductile low-carbon steel (0.2% C), or a stainless steel (18% Cr, 8% Ni), or a high-speed cutting steel (18% W, 5% Cr, 0.7% C). The extremely high variety of steels manufactured is mainly derived from pig iron by a process that involves the removal of impurities in the pig iron by oxidation and then adding alloying elements as required.

The majority of this so-called primary steel is now made using the oxygen furnace in which high-pressure oxygen is blown into a mixture of scrap and molten pig iron for about 15 minutes, to produce about

Table 9.1 Energy needed to produce 1 tonne of steel

| | Electric-arc furnace (100% scrap) | Blast furnace | |
		Oxygen furnace (30% scrap)	Open-hearth furnace (40% scrap)
total energy input (MJ/tonne steel)	13 000	22 000	25 000
% energy efficiency*	55	35	30

$$*\text{Energy efficiency} = \frac{\text{theoretical energy to convert ore to steel}}{\text{actual quantity of energy used}}$$

300 tonnes of steel in 30–40 minutes. As in the blast furnace, the impurities are removed in a slag formed with added lime. The oxidation of the impurities is highly exothermic and the scrap is added to absorb the excess heat. Because of its rapid turn-round time and more efficient use of energy, the oxygen furnace has replaced the open-hearth furnace that was the preferred method until about 20 years ago. The oxygen furnace produces large quantities of fine red dust, Fe_2O_3, and efficient fume- and dust-extraction systems have had to be developed to reduce this pollution problem.

Electric-arc furnaces are used to produce secondary steel, i.e. steel made from steel scrap as opposed to pig iron. They produce 25–125 t of steel in 2–4 hours and are used to make high-grade special steels that justify the high cost of electricity and pollution-control equipment. These costs have to be incorporated into the selling price. Because electric-arc furnaces use scrap, their total energy costs are lower than for other processes (Table 9.1).

The iron and steel industry is ranked as the second most polluting industry in the UK – coke production is number 1 – and the cost of adequate pollution-control measures is high because of the large quantities of materials handled. In a new steelworks these costs are likely to be a quarter or a third of the total capital costs. As the steel-making process is no more efficient after the installation of the control equipment, this added expenditure, plus the running costs, makes steel production more expensive in countries with strict pollution-abatement policies.

9.3 Corrosion

The Eh–pH diagram for Fe–H_2O (Figure 9.3) indicates that iron is unstable under the conditions existing on the Earth's surface. This instability is exhibited by the formation of rust (a hydrated iron(III) oxide of variable composition), which is the best-known example of **corrosion** – the deterioration of a substance by environmental constituents. The costs of corrosion are enormous, having been estimated as costing the UK 3.5% of its gross national product.

The corrosion process involves the transfer of electrons as the corroding metal is oxidized (Eqn 9.14). This means that for corrosion to occur there must be an electron acceptor (usually oxygen), but it need not be present at the site of corrosion if the electrons can be transferred to it at some other site:

$$M \xrightarrow{\hspace{1cm}} M^{2+} + 2e^- \xrightarrow{\hspace{0.3cm}O\hspace{0.3cm}} O^{2-} \xrightarrow{\hspace{0.3cm}M^{2+}\hspace{0.3cm}} MO \qquad (9.14)$$

The presence of water often enhances the rate and extent of corrosion:

(a) it can provide electron acceptors in either acid (Eqn 9.15) or alkaline solutions (Eqn 9.16):

$$M \xrightarrow{\hspace{1cm}} M^{2+} + 2e^- \xrightarrow{\hspace{0.3cm}2H^+\hspace{0.3cm}} H_2 \qquad (9.15)$$

$$2M \xrightarrow{\hspace{1cm}} 2M^{2+} + 4e^- \xrightarrow{\hspace{0.3cm}O_2 + 2H_2O\hspace{0.3cm}} 4OH^- \qquad (9.16)$$

(b) it can provide an electron-transfer medium, particularly if it contains dissolved ions (as do sea water, or water plus road de-icers);

(c) it can remove reaction products that could inhibit further reaction by building up a protective coating, especially if it contains dissolved ions that can form soluble complexes with the metal ions, e.g. $[FeCl_4]^-$;

(d) it can modify the structure of the reaction products so that they provide less inhibition to corrosion – e.g. rust is hydrated iron(III) oxide, $Fe_2O_3 . nH_2O$, whose volume is much greater than that of metallic iron: rust forms a loose, porous coating, whereas in the *absence* of water Fe_2O_3 will form a coherent, non-porous protective layer because its volume and crystal structure are much closer to that of metallic iron.

The surprising feature about corrosion is not that it occurs, but that in many cases the extent is much less than we might expect from diagrams such as Figure 9.3 and from a knowledge of the free energies of formation, ΔG_f^0, of the oxides (Table 9.2). The oxides all appear to be very much more stable than the metals, yet nickel and chromium are added to iron to form stainless (non-corroding) steel, and aluminium

(a) Paint

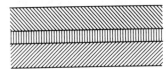

topcoat – impervious to O_2 and water

primer + undercoats – bind top coat to metal

metal

(b) Oxide coating

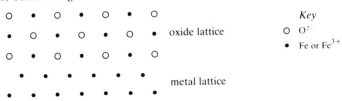

oxide lattice

metal lattice

Key

O O^{2-}

● Fe or Fe^{3+}

(c) Galvanization

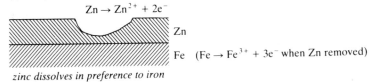

$Zn \rightarrow Zn^{2+} + 2e^-$

Zn

Fe $(Fe \rightarrow Fe^{3+} + 3e^-$ when Zn removed$)$

zinc dissolves in preference to iron

(d) Sacrificial anode

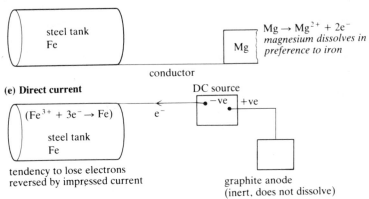

steel tank
Fe

conductor

$Mg \rightarrow Mg^{2+} + 2e^-$
magnesium dissolves in preference to iron

Mg

(e) Direct current

$(Fe^{3+} + 3e^- \rightarrow Fe)$

steel tank
Fe

e^-

DC source

−ve +ve

graphite anode
(inert, does not dissolve)

tendency to lose electrons
reversed by impressed current

Figure 9.9 A number of methods that may be used to prevent the corrosion of iron.

Table 9.2 Free energies of formation for the oxides of some metals

Metal oxide	Al_2O_3	Fe_2O_3	MgO	NiO	Cr_2O_3
$\Delta G_f^0 (kJ\,mol^{-1})$	−1580	−740	−570	−210	−1050

is widely used as an untreated construction material. In each case the oxide has a crystal structure closely related to that of the metal. A thin, non-porous coating of oxide rapidly forms on the metal's surface preventing further contact between unreacted metal and oxygen, or other electron acceptor, and further corrosion is prevented. Prevention of contact between reactants is one way of stopping corrosion; the second method is to provide excess electrons to react with the electron acceptor in place of the metal.

The protective coating may be achieved by adding a simple physical barrier (Figure 9.9a) such as paint or oil, or it may involve a chemical change of the metal surface. An oxide coating can be produced on iron by treating it with oxidizing agents like dichromate, $Cr_2O_7^{2-}$ (Figure 9.9b); a phosphate coating such as iron(III) phosphate, $FePO_4$, results from treatment of the iron with phosphoric acid, H_3PO_4. The provision of extra electrons may be achieved:

(a) by coating iron with a less electronegative element – zinc is used to galvanize steel (Figure 9.9c), the zinc being removed in preference to the iron;

(b) by connecting the iron to a piece of less electronegative element – the so-called sacrificial anode, e.g. magnesium or zinc, which is replaced when dissolved: this technique is used with storage tanks and ships (Figure 9.9d);

(c) by applying to the iron a current from the negative terminal of a low-voltage DC source, the positive terminal being connected to a non-sacrificial, inert anode (Figure 9.9e) – this technique is used on pipelines, ships, tanks, and even on the Panama Canal lock gates.

Recently microbial corrosion has been recognized as a major problem under reducing conditions in which sulphur compounds are available to act as electron acceptors (Eqn 9.17).

$$M + SO_4^{2-} + 8H^+ \xrightarrow{\text{bacteria}} \frac{8}{n}M^{n+} + S^{2-} + 4H_2O \qquad (9.17)$$

This problem can occur either externally in soils and sediments, or internally in storage tanks and pipelines with water–oil mixtures. Methods of protection are the same as for aerobic corrosion.

10 Aluminium

Aluminium Abundance by weight (the relative abundance is given in parentheses): Earth, 1.8% (8); crust, 8.2% (3); ocean, 1 ppb (31).

Aluminium is the third most abundant element in the Earth's crust and, like oxygen and silicon, it is widely dispersed in silicate minerals. In the cases such as feldspars and clays, in which the aluminium forms an essential structural component, the minerals are described as aluminosilicates. In other minerals, such as pyroxenes and amphiboles, the aluminium is only present as an isomorphous substituent for varying proportions of silicon: its presence is not essential for the stability of the crystal structure. The larger size of the aluminium compared to silicon allows it to form stable octahedral structures (Figure 10.1) with oxide ions, $[AlO_6]$, as well as tetrahedral units. In the clay minerals (Chapter 12), the aluminosilicates consist of layers of $[AlO_6]$ octahedral units and $[SiO_4]$ tetrahedral units. In both tetrahedral $[AlO_4]$ and octahedral $[AlO_6]$ units there is a considerable degree of covalent bonding between the aluminium and the oxygen.

The proportion of aluminium in weathered material gradually increases as weathering proceeds. This is because the solubility of aluminium is lower than that of silicon in the pH range 4.5–10. The changes occurring during weathering therefore follow the order shown in Figure 10.2, unless acid conditions prevail. The proportion of aluminium in the soil can be used to indicate either how intensely or for how long weathering has continued. The build-up of iron oxides will also occur at a pH greater than 4.5, unless reducing conditions, due to waterlogging, say, aid their removal. In tropical climates weathering to form insoluble layers of aluminium and iron oxides can be very rapid. If they are large enough, these aluminium-rich bauxite deposits can be used as the starting material for the production of aluminium.

10.1 Aluminium in industrial systems

Aluminium is second only to iron in its use as a metal (18×10^6 t, 1989). The aluminosilicates and the hydrated oxide are both

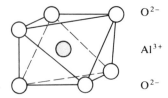

if the oxygens are close-packed, the octahedral hole
has a radius of 102 pm

Figure 10.1   Octahedral structure involving aluminium and oxygen, and found in a
variety of minerals.

increasing degree of weathering

→

aluminium in three-dimensional aluminosilicates e.g. feldspars	clays		aluminium oxides
	2 : 1 clays	1 : 1 clays	e.g. gibbsite
$NaAlSi_3O_4$	$Al_4(Si_4O_{10})_2$ $(OH)_4$	$Al_4Si_4O_{10}$ $(OH)_8$	$Al_2O_3 \cdot 3H_2O$
Si/Al 3/1	2/1	1/1	0/1

iron in silicates ⟶ goethite $FeO.OH$ or haematite Fe_2O_3

Figure 10.2   Chemical changes that occur as weathering progresses.

particularly stable, whereas the metal is even less stable than metallic
iron under the conditions existing at the Earth's surface. Therefore the
production of pure aluminium is a very energy-intensive process that
follows a quite different route from that taken in producing iron and
steel.

Chemically the problem is to find a method of reducing the
combined aluminium(III) to free elemental aluminium (0), as the free
energy of formation (ΔG_f^0) for Al_2O_3 is $-1580\,kJ\,mol^{-1}$, whereas
ΔG_f^0 for Fe_2O_3 is $-740\,kJ\,mol^{-1}$. Carbon and carbon monoxide will
not reduce Al_2O_3 at industrially feasible temperatures (less than
$2000\,°C$). Reduction involves the addition of electrons to the species
being reduced and an electric current can be used as the source of
electrons, providing its potential (voltage) is high enough.

The standard free energy, ΔG^0, and the standard electrode potential, E^0, are related (Eqn 10.1). The more negative the free energy of

formation, the greater the amount of electrical energy that is going to be required.

$$\Delta G^0 = -nE^0F \tag{10.1}$$

where F is the Faraday constant ($96\,487\,C\,mol^{-1}$).

The production of electricity is generally an inefficient energy-conversion process, especially when fossil fuels are used in power stations. By the time losses during generation and transmission have been taken into account, only 20–30% of the energy in the fuel is actually supplied to the consumer. Aluminium was initially produced where there was abundant hydroelectricity and good harbours through which to bring in the bauxite ore. In the 25 year period from 1950, smelters using fossil-fuelled power stations were built close to the markets, but smelters are now being built in the bauxite-producing countries such as Australia and Brazil where there are cheap power supplies, or on sites similar to those used initially – those with good harbours and hydroelectricity, such as Canada. The production of 1 tonne of primary aluminium requires more than four times as much energy as 1 tonne of copper or zinc, and seven times as much energy as 1 tonne of raw steel.

Bauxite is mined, usually by open-pit methods, from deposits formed in tropical conditions. The crushed rock is treated with hot sodium hydroxide under pressure to remove impurities, and pure Al_2O_3 is produced. The alumina, Al_2O_3, is reduced to molten aluminium in large electrolytic cells. Aluminium chloride ($AlCl_3$) cannot be used because, unlike sodium chloride, it is largely covalent and sublimes at 180 °C.

The identified world reserves of bauxite have risen more than fourfold since 1965, with most of the increase occurring in Australia and Guinea, so there is no immediate prospect of shortages of raw materials. The recycling of aluminium would give great savings in energy as remelting and reprocessing would require about $10\,GJ\,t^{-1}$ rather than the $260\,GJ\,t^{-1}$ used in primary aluminium production. The major problem with recycling is the separation of the required element from the other components in the manufactured article. The properties of aluminium are very markedly altered by small amounts of other metals or silicon (formed from glass, Eqn 10.2) that alloy with it on remelting.

$$Na_2SiO_3 + CaSiO_3 \xrightarrow{\text{Al + heat}} 2NaAlO_2 + Ca(AlO_2)_2 + 2Si \tag{10.2}$$

Aluminium is non-magnetic and the metal is not easily separated from contaminants, but unless it remains as the metal during the separation stage the energy advantage of using recycled aluminium is lost.

Aluminium has the properties of lightness, good thermal and electrical conductivity, workability, resistance to corrosion and tensile strength when alloyed. It is used in the construction, transport, electrical and consumer-durables industries, and for containers and packaging.

The corrosion-resistance properties of aluminium arise from the coherent coating of Al_2O_3 that rapidly forms on the surface of the metal and prevents further oxidation occurring (Eqn 10.3).

$$Al \xrightarrow[\substack{or \\ corrosion}]{oxidation} Al^{3+} + 3e^- \quad \text{removed by electron acceptor} \qquad (10.3)$$

The solid Al_2O_3 coating is a very poor conductor of electrons, has a low solubility, and has a crystal structure with few defects that is very closely related to that of aluminium metal. Once the coating has formed there is no further contact between the metal and the electron acceptor, O_2, and there is no possibility of electron transfer occurring through the insulating layer of Al_2O_3. However, under either alkaline (pH > 10) or acid (pH < 4.5) conditions, the protective coating will be removed because soluble species are formed; 'corrosion' will then occur. Similarly sea water contains a high concentration of chloride ions, Cl^-, that form soluble complexes with aluminium, and this too destroys the protective coating.

10.2 Aluminium in solution

The presence of aluminium species appears to play a part in controlling the pH of acid soil waters in the range pH 3–6 (Figure 10.3). Increasing inputs of acid, whether due to acid rainfall or to the decaying organic matter from forestry, will cause increasing removal of aluminium from soils that have a low buffering capacity. Soluble aluminium is moderately toxic to most plants and the effects of acidification of soils are to reduce the concentration of some essential elements and increase the concentration of toxic elements. It has been suggested that there are three phases in the effects of acid rain on forested areas.

(a) The trees benefit from the increased sulphur and nitrogen added to the poor soils.
(b) Continued acidification reduces the soil's neutralizing, or buffering, capacity and leaches out soluble metallic nutrients such as calcium and magnesium.

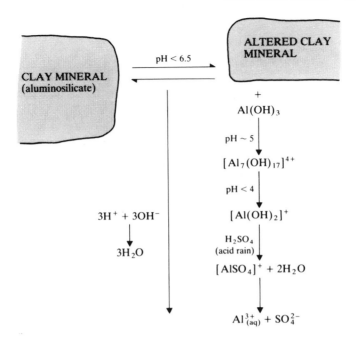

Figure 10.3 A possible reaction sequence for the buffering of hydrogen ions in acid soils by reaction with aluminium species. (After Ulrich, B. 1980. Production and consumption of hydrogen ions in the ecosphere. In *Effects of acid precipitation on terrestrial ecosystems*. Hutchinson, T. C. and M. Havas (Eds), 255–82. New York: Plenum Press.)

(c) The process accelerates and aluminium is extracted from the soil minerals, making the soil toxic to the trees. At the same time heavy metals, also toxic, are mobilized. The critical pH appears to be about 4.2, with aluminium as the main killer below this value.

Aluminium is less toxic to humans at low concentrations and the quantities of dissolved aluminium in water are normally very low ($50\ \mu g\ dm^{-3}$). Water collected in reservoirs from upland catchments is often highly coloured by organic compounds. This colour, together with any other suspended colloidal matter, can be removed by the addition of sodium aluminate ($NaAlO_2$) or aluminium sulphate ($Al_2(SO_4)_3$, 'alum'). These soluble aluminium compounds are hydrolysed in water (Eqn 10.4) and converted to aluminium hydroxide ($Al(OH)_3$), a gelatinous precipitate with a high surface area which helps to remove the colour and colloids when the suspension is filtered.

$$Al^{3+} + 3HCO_3^- \rightleftharpoons Al(OH)_3 + 3CO_2 \qquad (10.4)$$

body fluid

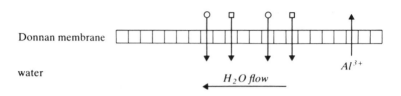

Donnan membrane

water

Al^{3+}

H_2O *flow*

the Donnan membrane prevents the large organic molecules passing through, but smaller species can penetrate the membrane – there is a net transfer of small species from regions of higher concentration to those of lower concentration

Key

large organic molecules, e.g. proteins

▢   ○   small organic molecules, the products of metabolism, e.g. urea, creatinine

Figure 10.4   The principle of the dialysis process used in haemodialysis of patients with defective kidneys.

Some of the added aluminium ions remain in solution and this has caused problems to kidney patients on dialysis machines. The **Donnan membranes** used in dialysis units (Figure 10.4) do not exclude aluminium as efficiently as do kidneys. The resulting higher intake of aluminium can eventually cause death, and this was a major problem until the cause was realized. The aluminium can be removed from the water used in the dialysis unit by passing it through an ion-exchange column, but because of the large volumes involved there is a significant increase in the cost of the treatment.

There has been a lot of speculation about possible links between aluminium and Alzheimer's disease. People suffering from Alzheimer's disease have been found to have higher than normal amounts of aluminium in their brains. Alzheimer's disease produces many of the symptoms of dementia shown by dialysis patients who have suffered from aluminium poisoning. However, it has not been possible to pinpoint whether aluminium is a causative agent or whether it just happens to accumulate because of some change in the biochemistry of Alzheimer's patients that has been brought about by some other agent. Concern about the possible chronic effects of low dosages of aluminium is bringing about a reduction in the use of aluminium compounds in water treatment. One alternative is to use iron(II) compounds which react with oxygenated water in a similar manner to aluminium (see Chapter 9), but are less efficient. Organic polyelectrolytes have also been used, but they can cause problems by introducing traces of potentially carcinogenic compounds into the water.

11 Calcium and magnesium

Magnesium Abundance by weight (the relative abundance is given in parentheses): Earth, 13.5% (4); crust, 2.1% (8); ocean, $1.3\,g\,dm^{-3}$ (5).

Calcium Abundance by weight (the relative abundance is given in parentheses): Earth, 2.1% (7); crust, 3.6% (5); ocean, $0.41\,g\,dm^{-3}$ (7).

Magnesium and calcium are the first two members of the alkaline-earth group of metals. They are both major components of the Earth's crust, and, because of their similar electronic structures (two electrons in the outermost shell), their chemical reactions are similar. Together with iron, they are the major cations found in basic igneous rocks.

Despite their chemical similarity, there is only limited substitution between magnesium and calcium in their minerals. This is because the ionic radius of Mg^{2+} is 65 pm, whereas that of Ca^{2+} is 100 pm. This size difference is sufficient to prevent the large-scale replacement of one ion by the other. It is iron (II), Fe^{2+} (radius 61 pm), that replaces Mg^{2+}; and sodium, Na^+ (radius 102 pm), that exchanges with Ca^{2+}.

The geochemical cycles of the two elements (Figure 11.1) are similar, except for the variations brought about by the differences in solubility between some of their compounds. The lower solubility of calcium carbonate ($CaCO_3$) compared to magnesium carbonate ($MgCO_3$) leads to extensive sedimentary deposits of limestone ($CaCO_3$) being formed from the dissolved calcium that has been transported by rivers into the ocean. Some magnesium (about 5% on average) was incorporated in these limestone deposits as they were precipitated. In addition, dolomite ($CaCO_3 \cdot MgCO_3$) may be formed as a mineral in its own right. Both calcium, as the sulphates anhydrite ($CaSO_4$) and gypsum ($CaSO_4 \cdot 2H_2O$), and magnesium, as the chloride ($MgCl_2 \cdot 6H_2O$) and as sulphate ($MgSO_4$), are found in evaporite deposits. Again the lower solubility of the calcium sulphates leads to them being formed first. These precipitation processes all favour the removal of calcium in preference to magnesium from sea water. There is a marked reduction in the proportion of calcium relative to magnesium in sea water ($Ca^{2+} = 1.2\%$ dissolved solids; $Mg^{2+} = 3.7\%$ dissolved solids) compared with river water ($Ca^{2+} = 20\%$ and $Mg^{2+} = 3.4\%$ of dissolved solids). The residence times in the ocean of

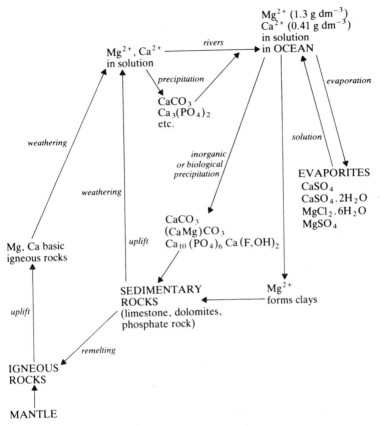

Figure 11.1 Geochemical cycles of magnesium and calcium.

magnesium (15×10^6 years) and calcium (1×10^6 years) also reflect the lower solubility of the calcium compounds. The reaction that removes the excess of the magnesium involves absorption by clay minerals and the formation of new clay minerals such as chlorites (Eqn 11.1).

$$Al_2Si_2O_5(OH)_4 + 5Mg^{2+} + SiO_2 + 10HCO_3^- \longrightarrow$$
$$\underset{\text{kaolinite}}{} \qquad \underset{\text{chlorite}}{Mg_5Al_2Si_3O_{10}(OH)_8} + 10CO_2 + 3H_2O \qquad (11.1)$$

11.1 Formation of calcareous rocks

The behaviour of calcium in the oceans underwent a major change with the development of life-forms that incorporated Ca^{2+} in their skeletons. Originally the removal of calcium carbonate from the oceans was by means of normal *inorganic* precipitation reactions (Eqn 11.2) which occurred when the combined concentration of calcium and carbonate ions exceeded the solubility product of calcium carbonate.

$$Ca^{2+}_{(aq)} + CO^{2-}_{3(aq)} \rightleftharpoons CaCO_{3(s)} \qquad (11.2)$$

The majority of limestones have been *organically* formed for, at least, the past 600 million years. The low solubilities of many inorganic calcium compounds, especially calcium carbonate and the apatites (e.g. hydroxyapatite, $Ca_3(PO_4)_2 \cdot Ca(OH)_2$; fluoroapatite, $Ca_3(PO_4)_2 \cdot CaF_2$), have made them especially suitable for skeleton building. Skeletons are either exterior to the soft parts – exoskeletons – as with insects and shellfish, or internal – endoskeletons – with the muscles and other soft tissues surrounding this skeleton, as in mammals and fish. The exoskeletons of insects are usually composed of organic polymers, such as chitin, and some sponges have silica (SiO_2) skeletons; but the majority of other invertebrates (with shells) and vertebrates (with internal skeletons) have calcium-containing hard parts.

The solubility of calcium carbonate is controlled by the concentration of both $Ca^{2+}_{(aq)}$ and $CO^{2-}_{3(aq)}$. Because sea water is a relatively concentrated solution containing a large number of components, species other than the free dissolved calcium and carbonate ions are present. Of particular importance are species called **ion pairs**. Ion pairs are associations between two oppositely charged ions which form new soluble species (Figure 11.2) and which therefore reduce the

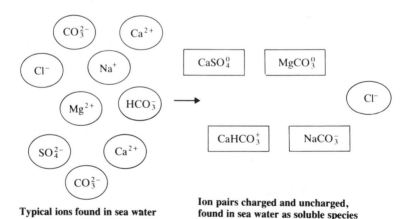

Typical ions found in sea water

Ion pairs charged and uncharged, found in sea water as soluble species

Figure 11.2 Examples of ion pairs found in sea water.

concentrations of the free hydrated ions. In addition, more complex species may be produced by the association of the metal ions with several anionic groups, e.g. $HgCl_4^{2-}$, the tetrachloromercury(II)ate ion. The concentration of free carbonate ion, $CO_{3(aq)}^{2-}$, is reduced to about 10% of the expected value due to formation of ion pairs: 60–70% of the carbonate is found as the $MgCO_3^0$ ion pair and 15–20% as $NaCO_3^-$. A greater proportion of the calcium (up to 90%) appears to be present as the free ion, $Ca_{(aq)}^{2+}$, with $CaSO_4^0, CaHCO_3^+$ and inorganic calcium complexes as the other major species. Theoretical calculations of compound formation based upon thermodynamic stability data can only be approximate because of the many interactions and our very incomplete understanding of what species are actually present in sea water.

In general, conditions that favour the extensive production of limestone deposits require clear waters and the absence of silicates brought into the area by rivers. As a consequence, many limestones are relatively pure, with some being at least 98% $CaCO_3$. These pure deposits are now much prized as sources of lime (calcium oxide, CaO) for industrial processes such as steel making, glass making and sugar refining. Millions of tonnes of lime are consumed in these industries each year. Unfortunately, in the UK, the areas of purest limestone are also areas of outstanding natural beauty and often in national parks. There is thus a conflict between the industrial requirements that lead to large quarrying operations and the need to preserve the few undeveloped areas left.

11.2 Water hardness

Carbonate rocks are rapidly eroded when exposed to water containing dissolved carbon dioxide (Eqns 11.3 and 11.4), though dolomite ($CaCO_3 \cdot MgCO_3$) is more resistant than limestone ($CaCO_3$).

$$CaCO_{3(s)} + CO_2 + H_2O \rightleftharpoons Ca_{(aq)}^{2+} + 2HCO_3^- \qquad (11.3)$$

$$MgCO_{3(s)} + CO_2 + H_2O \rightleftharpoons Mg_{(aq)}^{2+} + 2HCO_3^- \qquad (11.4)$$

The higher the concentration of dissolved calcium and magnesium in the water, the harder the water is said to be. **Hard water** containing the bicarbonates of calcium and magnesium is said to have **temporary hardness** because the equilibria in Equations 11.3 and 11.4 are reversed on boiling, and the carbonates are precipitated. This removes the calcium and magnesium from solution. If the precipitation occurs in boiler tubes, kettles or hot-water pipes, the efficiency of heat transfer is reduced, the tubes may become blocked and the operating costs are increased. A less common type of hardness in fresh waters is due to the presence of sulphate ions and is called **permanent hardness**: it is not all removed on boiling the water. The solubility of calcium sulphate is lowered with a rise in temperature so there is a slow build-up of deposits in the pipes. Both types of hardness lead to the formation of insoluble scums with soaps.

A **soap** is the soluble sodium or potassium salt of a long-chain fatty acid, such as stearic acid (Figure 11.3). The calcium and magnesium salts are insoluble and do not have the dirt-removing properties of the sodium and potassium salts. Therefore their formation increases the amount of soap that must be used and the scum may be difficult to remove from washed clothes. Organically based detergents usually contain sulphonic acid groups (Figure 11.3) and form soluble salts with calcium and magnesium. Therefore there is no scum and their effectiveness as cleaning agents is only slightly reduced. The cleaning action is due to the hydrophobic ('water hating') organic tail of the

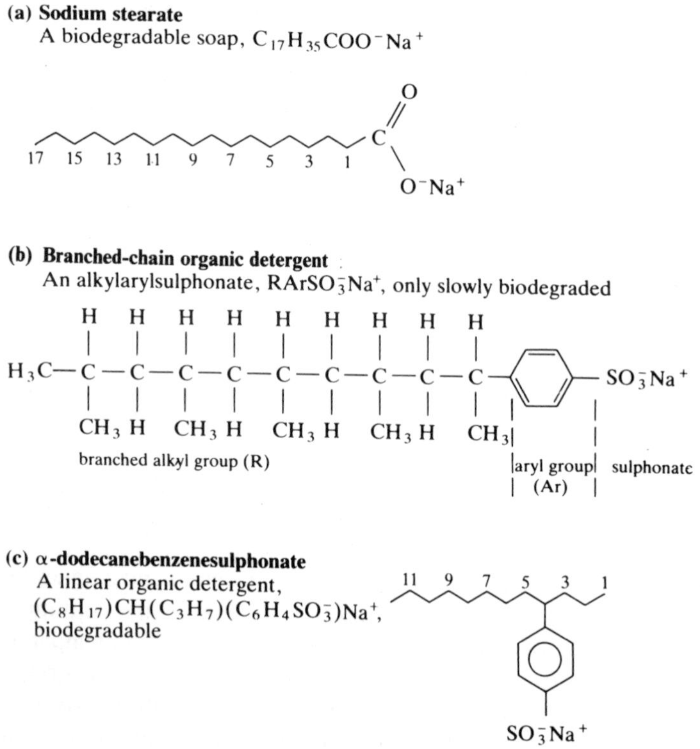
(a) Sodium stearate
A biodegradable soap, C17H35COO−Na+
O
C
17 15 13 11 9 7 5 3 1
O−Na+
(b) Branched-chain organic detergent
An alkylarylsulphonate, RArSO3−Na+, only slowly biodegraded
H H H H H H H H H
H3C—C—C—C—C—C—C—C—C—C—
SO3−Na+
CH3 H CH3 H CH3 H CH3 H CH3
branched alkyl group (R)
aryl group (Ar)
sulphonate
(c) α-dodecanebenzenesulphonate
A linear organic detergent,
(C8H17)CH(C3H7)(C6H4SO3−)Na+,
biodegradable
11 9 7 5 3 1
SO3−Na+

Figure 11.3 Examples of soaps and synthetic detergents.

molecule being attached to the dirt or grease particle, and the hydrophylic ('water loving') portion of the molecule forming hydrogen bonds with water molecules. This allows the dirt particles to be dispersed in the water and removed from the object being cleaned. Soaps are **biodegradable** (broken down by micro-organisms); excess soap entering sewers is rapidly converted to the insoluble calcium and magnesium salts that do not lower the surface tension of water and so do not cause foaming. The original organic detergents were branched-chain compounds (Figure 11.3) that were only very slowly biodegraded: these caused extensive foaming in sewage works and water systems. Straight-chain compounds (Figure 11.3) that are rapidly biodegradable are now used to reduce these problems.

Detergents contain other compounds – up to 90% of solid detergents are 'other compounds' – that may be viewed with some concern as regards their environmental impact. One group of additives that can cause problems are the polyphosphates. These were discussed in Chapter 7.

Though hard water causes problems due to 'furring' of pipes and kettles and the inefficient use of soaps, soft water can also cause problems. Soft water tends to have a low pH, due to the absence of dissolved carbonates, and it can be corrosive enough to dissolve the metal from water pipes. Lead is soluble under these conditions and lead poisoning is a recognized hazard in soft-water areas where lead pipes are used. In the UK, the water authorities artificially harden the water in soft-water areas. This may be achieved by mixing hard and soft waters, or by adding lime to the water supply to raise the pH.

11.3 Heart disease

A number of studies carried out in the past appear to have indicated a negative correlation between death (mortality) from heart disease and the hardness of the water supply, i.e. the harder the water, the fewer the deaths from heart disease. However, it has not been possible to relate any specific 'water factors' directly to specific heart disease effects despite earlier views that such a factor would be identified. This section will illustrate the extreme difficulties that arise when attempting to relate possibly subtle environmental factors to the health of the population. Similar problems are being encountered with the relationship between high aluminium levels in the brain and Alzheimer's disease. Heart disease is the leading cause of death in industrial countries and it has been estimated that in England and Wales as many as 10 000 extra deaths per annum due to heart disease occur in men aged 45–64 years who live in soft-water rather than hard-water areas. The difficulty has been to determine whether there is a *real* connection between the water supply and heart disease. There could be some other factor affecting the incidence of heart diseases that also happens to correlate with water type.

In many countries there is a geographical distribution of areas of higher and lower incidences of heart disease that appears to correlate with particular geological and geochemical environments. In European countries, populations living on older rocks tend to have higher death rates due to heart attacks. For instance, in 1967 death rates per 100 000, for both sexes and for all ages, were 314 ± 29 for rocks older than 600 million years and 159 ± 50 for rocks younger than 180 million years. The older geological rocks tend to contain fewer carbonates and the water supply in these areas is therefore softer. Geochemically the rocks also have lower concentrations of trace elements. In the USA, higher heart-disease death rates are associated with rocks, soils and waters deficient in trace elements, but again the waters are softer. In 1968–72 death rates per 100 000 whites, for both sexes and for all ages, were 366 ± 32 (high in trace elements) and 428 ± 39 (low in trace elements) for heart diseases, but 432 ± 70 (high in trace elements) and 424 ± 44 (low in trace elements) for other causes of death. These results could indicate that it is the lack of some trace elements that is related to the higher death rates, rather than the presence of calcium and magnesium salts in the water supply. In that case the artificial hardening of the water supply by the addition of lime would not decrease death rates.

Adverse climatic conditions often correlate with death rates. In the UK the soft-water areas are in the west where the climate is wetter so the soft-water–mortality correlation may be accidental.

A further difficulty arises because heart disease is a general term including a number of clinically different conditions. In different countries the associations with water supplies have been between different members of the group, with no common pattern from country to country.

It is clear that many of the investigations have either not been planned carefully enough or suffered from a lack of adequate data. One study was originally interpreted as indicating a 47% reduction in mortality rates over a 10 year period when the hardness of the water was changed from 0.5 ppm to 220 ppm. Unfortunately, the population change that had occurred was not taken into account, and when the data was reinterpreted taking this factor into account the reduction in death rate was only 16% compared to a 13% reduction found for the whole of the USA during the same period.

Detailed studies of individuals and of the possible environmental factors that may be affecting them are extremely expensive: large populations are required to avoid statistical bias and the study must be carried out for many years. The alternative is to make use of data already collected, such as analyses of water supplies by water authorities and mortality data from death certificates, together with small-scale detailed surveys. Problems arise because (a) health authorities do not cover exactly the same areas as water authorities; (b) populations move and the places in which people die may not be the places in which they lived for most of their lives; (c) the death certificate may not accurately indicate the condition of the heart in relation to other health factors at the time of death; (d) social and economic factors may be difficult to identify.

The specific water factor has not been identified, though suggestions were that:

(a) hardness may provide protection – possibly by preventing dissolution of lead and cadmium from water pipes: both these metals can produce high blood pressure, one of the precursors to heart attacks;
(b) some of the trace elements in hard water may provide protection – e.g. lithium is thought to be beneficial, at small concentrations, in reducing anxiety;
(c) the low concentration of trace elements in soft water may result in a deficiency of protective agents.

12 Sodium and potassium

Sodium   Abundance by weight (the relative abundance is given in parentheses): Earth, 0.3% (9); crust, 2.8% (6); ocean, 1.05% (4).

Potassium Abundance by weight (the relative abundance is given in parentheses): Earth, 150 ppm (14); crust, 2.6% (7); ocean, 0.39 g dm^{-3} (8).

Most sodium compounds that are not silicates have high solubilities. This means that once the primary silicate minerals such as feldspars have been broken down by weathering, the sodium is rapidly leached out of the soils and transferred to the ocean. The only barrier to the rapid removal in solution is the tendency of sodium ions to be strongly adsorbed by clays and organic materials that have surface negative charges. The sodium concentration (mean: 6 mg kg^{-1}) in fresh waters is lower than that of calcium (mean: 15 mg kg^{-1}), a consequence of the greater stability of sodium silicates and their lower average concentration in the Earth's crust. In sea water the high solubility of sodium salts is indicated by sodium's long residence time (210×10^6a) and the fact that it is the major cation (10.54 g kg^{-1}), with magnesium second (1.27 g kg^{-1}), calcium third (0.4 g kg^{-1}), and potassium fourth (0.39 g kg^{-1}).

Potassium and sodium are in the same group in the periodic table: their simple salts have similar solubilities but their geochemical behaviour differs in a number of ways (Figure 12.1). Sodium and calcium undergo isomorphous replacement in silicates because of their similar ionic radii, whereas potassium is found in separate primary igneous minerals (orthoclase feldspar and micas). The weathered potassium silicates release potassium ions but these are even more strongly adsorbed by negatively charged clay and organic colloids than are sodium ions. Unlike sodium, the potassium is readily reincorporated into silicate structures with the formation of clay minerals and its concentration in biological material is about 15 times greater than that of sodium. As a consequence the concentration of potassium in fresh waters is about one-third that of sodium and its residence time in the ocean is 10 million years.

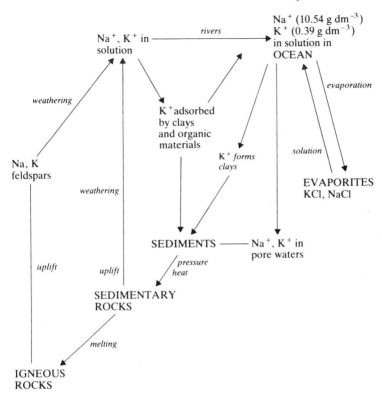

Figure 12.1 Geochemical cycles of sodium and potassium.

12.1 Clay minerals

The clay minerals are hydrous aluminium silicates (contain Al, Si, O and OH) of small size with a layered structure. This characteristic structural feature is made up of sheets of either tetrahedral $[SiO_4]$ units or octahedral $[AlO_6]$ units (Figure 12.2). The $[SiO_4]$ tetrahedra are linked together by the sharing of three basal oxygen atoms with their apexes all pointing in the same direction. The octahedral units contain two layers of close-packed oxygens and hydroxyls surrounding the aluminium. The relative number of oxygen and hydroxyl groups varies to satisfy the charge-balance criteria for the structure.

The simplest of the clay structures in shown by kaolinite $(Al_4Si_4O_{10}(OH)_8)$, with one octahedral sheet and one tetrahedral sheet forming a 1:1 clay mineral (Figure 12.2a). Successive 1:1 layers are stacked above each other and held together by hydrogen bonds between the oxygens in one layer and the hydroxyl groups in the next layer. The hydrogen bonds prevent other groups from entering between the individual layers and keep the structure relatively rigid. Kaolinite shows only a limited amount of isomorphous replacement of the Al and Si, aluminium and silicon, and the composition of kaolinite corresponds very closely to the ideal formula given above.

Members of the other major group of clay minerals, the smectites, each consist of one octahedral sheet sandwiched between two tetrahedral sheets (Figure 12.2b). Montmorillonite $(Al_4(Si_4O_{10})_2 (OH)_4)$ and illite $(K_{0-2}Al_4 (Si_{8-6}Al_{0-2}) O_{20}(OH)_4)$ are the two most common members of the 2:1 clays. In each case the oxygens of one 2:1 layer always face the oxygens of the next layer, and therefore no hydrogen

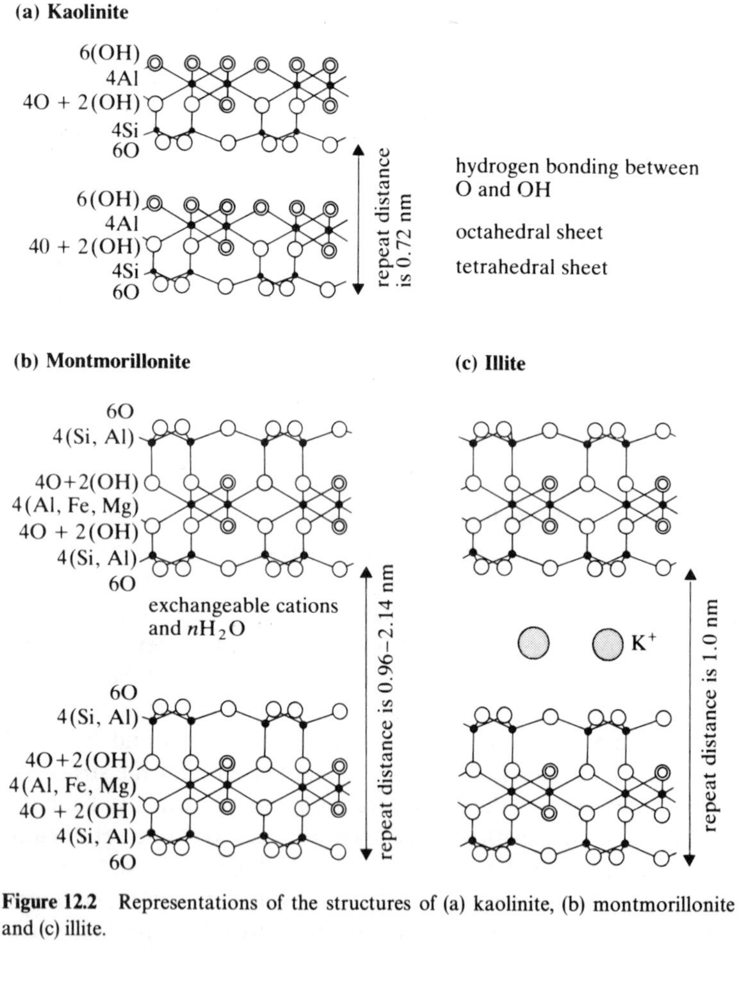

Figure 12.2   Representations of the structures of (a) kaolinite, (b) montmorillonite and (c) illite.

bonding can take place. The layers are not so strongly held together, and ions such as K^+ in illite and small molecules such as water in montmorillonite can enter between the layers. Montmorillonite-containing clays are described as expanding clays because when water enters between the layers the repeat distance (Figure 12.2b) can increase from 0.96 nm to 2.14 nm. During dry periods the water will leave the clay, and the clay will contract again. The entry of potassium between the layers to form illite is only reversed with difficulty. The potassium ions tend to hold the layers together at a fixed distance apart and make the entry of water difficult. Kaolinite and illite are described as non-expanding clays.

Montmorillonite clays never have their ideal formula and show extensive isomorphous substitution with up to 15% of the silicon in the tetrahedral units replaced by ions such as aluminium. In the octahedral units there may be up to 100% replacement of the aluminium by ions such as Mg^{2+} and Fe^{2+}, with smaller amounts of $Zn^{2+}, Ni^{2+}, Li^+, Cr^{3+}$, etc. The substitution of Al^{3+} for Si^{4+} and the divalent cations for Al^{3+} leaves a positive-charge deficiency that may be compensated for (a) by OH^- replacing O^{2-}, (b) by excess cations entering into the one-third unfilled octahedral positions found in the ideal structure, or (c) by sorption of cations on to the surface of the layers. All three compensation mechanisms occur and montmorillonite has a high cation-exchange capacity due to mechanism (c).

The ability of a soil to exchange cations is called the cation-exchange capacity (CEC) quoted as milli-equivalents per 100 g of soil. (1 milli-equivalent = 1 millimole of a *uni*positive cation, e.g. Na^+, which is 23×10^{-3} g; 1 milli-equivalent = 0.5 millimole of a *di*positive cation, e.g. Ca^{2+}, which is 20×10^{-3} g.) Organic materials exchange cations mainly due to the presence of the carboxylate group, $-COO^-$. Humus usually has a very high cation-exchange capacity, e.g. CEC of peat = 300–400 compared to CEC of typical soil = 10–30. Kaolinite has a relatively low CEC, which is due to the exchange of H^+ ions from the hydroxyl groups on the clay surfaces. The hydroxyl groups of montmorillonite react similarly, but the greater degree of isomorphous substitution and the consequent excess negative charges lead to a higher CEC.

An equilibrium state is set up in a soil in which the proportion of an ion in solution and sorbed on to the surface of the clay or humus depends on (a) the concentration of metal ions present, and (b) the acidity of the soil. The equilibrium can be changed by changing either (a) or (b) or both (Figure 12.3). The concentration of an ion in solution may be reduced by uptake by plant roots. Some ions will then be released from the cation-exchange sites to restore the equilibrium and

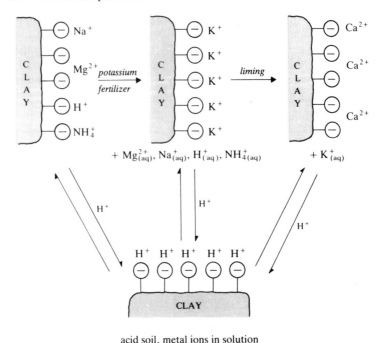

acid soil, metal ions in solution

Figure 12.3   Examples of how changing the concentration of cations can change the equilibrium between the cations held by the clay and those in solution.

hence more of the ion will become available to the plant. The process will continue until there is no more of that ion sorbed on the clay or humus. In general a high proportion of the cation-exchange sites on organic matter are occupied by hydrogen ions unless the pH of the soil is raised by liming. The cation-exchange sites can be an important reservoir of essential metal ions for plants as they reduce the rate of leaching of these soluble ions, provided that the pH is not too low.

12.2 Potassium fertilizers

Potassium is one of the three major fertilizer elements, the others being nitrogen and phosphorus. In 1939 about 67 000 t of potassium was used in fertilizers in the UK. Annual consumption is now about 350 000 t. In general the potassium status of UK soils is satisfactory and only maintenance dosages are added – enough potassium is added to compensate for the potassium removed in the crops. This is because any excess potassium added is largely retained in the soil by sorption on clays and organic matter. (This is unlike the case of nitrogen, in which any excess is leached out of the soil.) Soils generally have low anion-exchange capacities and once the nitrogen is converted to nitrate it can be readily removed in solution. The clay minerals have a limited anion-exchange ability through the replacement of hydroxyl ions on crystal edges (Eqn 12.1).

$$\text{clay–OH} + \text{anion}^- \longrightarrow \text{clay–anion} + \text{OH}^- \qquad (12.1)$$

Nitrate is only weakly held at these sites. Phosphate ions, PO_4^{3-}, are more strongly held and also react with aluminium and iron oxides to form insoluble compounds that prevent leaching out of the phosphate.

In areas where crops have been grown for many years without the addition of adequate potassium-containing fertilizers, yields gradually decrease as the potassium from between the illite layers is slowly removed. If potassium fertilizer is then added, the increase in yield is not as great as might be expected. This is because the potassium returns to the illite structure rather than remaining immediately available for plant growth. As a consequence the farmers are faced with the high costs of potassium fertilizers without receiving a comparable increase in crop yield. This constitutes a major problem for subsistence farmers, who cannot afford to carry the extra costs without an immediate return. As a result, unless external help is provided, they cannot afford to improve their impoverished soils. High yields of any crop can be sustained only by replacing the nutrients removed with the crop. This is expensive enough for the farmers of the developed countries, even with high food prices and subsidies; applying adequate fertilizer is virtually impossible for the farmers of Third World countries who lack the same organizational back-up. Large areas of the Earth's surface are less productive than they might be with proper investment and management.

12.3 Fluids in organisms

A cell absorbs required chemical species through the phospholipid membrane that constitutes the cell wall and excretes unwanted material through the same barrier. In the case of unicellular organisms the interchange is directly with the environment in which it lives, but with more complicated organisms the majority of the cells are not in contact with the external environment. The development of multicellular organisms has been paralleled by the development of a body-fluid system that maintains a controlled internal environment, transfers essential chemicals to the individual cells and removes the unwanted waste chemicals. The extracellular fluid is usually mainly composed of an aqueous solution containing sodium, potassium, magnesium and calcium cations, with chloride and bicarbonate as the major anions, together with soluble proteins and organic acids. Other soluble species are present as minor constituents.

The concentrations of ions inside and outside cells are often quite different. Particularly marked in humans are the high concentrations of potassium and magnesium ions inside cells and the high concentrations of sodium and calcium ions in the blood serum (Table 12.1).

Maintaining these segregations is of great importance for the correct metabolic functioning of organisms. For example, the difference in sodium and potassium concentrations across cell membranes leads to electrical potential differences across the membrane. Changes in this potential difference by the sudden influx of sodium ions enable the transmission of nerve impulses. Both sodium and potassium are able to pass through the cell wall. Because of the concentration differences, the natural tendency would be for the sodium to pass from the high-concentration solution outside the cell to the lower-concentration solution inside the cell, and for potassium to move in the opposite direction (Figure 12.4). The maintenance of the Na^+–K^+ differential across the cell wall involves the active removal of sodium from the interior of the cell by an ion-pump mechanism. It has been suggested that up to half the basal metabolic rate in humans is dedicated to operating this sodium pump, which is driven by the hydrolysis of ATP.

The walls of cells act as Donnan membranes. This means that some of the dissolved species are unable to pass through the cell wall or can only pass through under certain conditions. Water, the solvent, passes through a semipermeable membrane from the more dilute solution to the more concentrated solution until the concentration of dissolved species is the same in both solutions. This process is called **osmosis**. The flow of solvent can be prevented by applying a pressure to the more concentrated solution. The size of the applied pressure required to prevent the solvent flow is called the osmotic pressure. The greater the concentration difference across the semipermeable membrane,

Table 12.1   Concentrations of cations in human intracellular and extracellular fluids

	Na^+	K^+	Mg^{2+}	Ca^{2+}
blood serum ($g\,dm^{-3}$)	3.3	0.2	0.03	0.1
cellular fluid ($g\,dm^{-3}$)	0.23	4.3	0.49	0.000 04

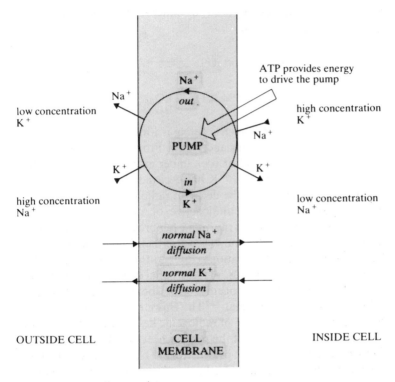

Figure 12.4   The transport of sodium and potassium ions across cell membranes by the sodium-pump mechanism.

the greater is the tendency for the solvent to pass through and the higher is the osmotic pressure. If the ionic strength inside a cell is greater than the ionic strength in the extracellular fluid, water will tend to enter the cell causing it to swell. Single-celled marine organisms placed in fresh water (which has a low ionic strength) are no longer in ionic balance with their surroundings; water enters the cells and eventually they burst. Conversely, freshwater organisms when placed in sea water (which has a high ionic strength) lose water and shrivel up as they dehydrate. Multi-celled organisms have developed quite sophisticated mechanisms to equalize ionic strengths and overcome osmotic effects either inside their bodies or due to external influences.

12.4 Radioactive potassium

The naturally occurring isotopes of potassium are $^{39}_{19}K$ (93.1%), $^{40}_{19}K$ (0.0119%) and $^{41}_{19}K$ (6.9%). ^{40}K is radioactive with a half-life of

1.27×10^9 a and undergoing decay by either K-electron capture or β^- decay (Eqns 12.2 and 12.3).

$$\underset{K \text{ electron}}{^{40}_{19}\text{K} + {}^{0}_{-1}\text{e}} \xrightarrow{\ (11\%)\ } \underset{\text{excited}}{^{40}_{18}\text{Ar}} \longrightarrow \underset{\text{ground state}}{^{40}_{18}\text{Ar}} + \underset{\text{gamma ray}}{\gamma} \qquad (12.2)$$

$$^{40}_{19}\text{K} \xrightarrow{\ (89\%)\ } {}^{40}_{20}\text{Ca} + \underset{\text{beta radiation}}{\beta^- \text{ (or } {}^{0}_{-1}\text{e)}} \qquad (12.3)$$

Potassium–argon **radiometric dating** is widely used for estimating the age of geological systems and events. Providing the ^{40}Ar has not diffused out of the mineral or rock under examination, the amount present indicates how much ^{40}K has decayed.

The energy released by the decay of radioactive ^{40}K is the third most important source of radioactive or radiogenic heat in the Earth's crust. Uranium and thorium each provide about three times as much heat as does ^{40}K now, because their longer half-lives have kept their present concentrations closer to their original concentrations. There is only about 10% of the original quantity of ^{40}K present in the Earth's crust now, together with about 50% of the original ^{238}U and 80% of the original ^{232}Th. This means that the ^{40}K must have been the major source of radiogenic heat in the early stages of the Earth's history. Most of the present heat flow through the crust can be accounted for by the presence of ^{238}U, ^{232}Th and ^{40}K in the crustal rocks. The natural background radiation on the Earth's surface is also largely due to the presence of these radioactive elements and their decay products.

Part Four

Minor elements and environmental problems

In the preceding chapters we have examined elements that are major components of the Earth and its life-forms. In those that follow we shall look at a number of elements that are generally present only in small quantities in natural systems. However, because of industrial and agricultural processes, the geochemical cycles of these trace elements have been significantly modified, leading to concern about their polluting effects. There is also a chapter looking at a number of persistent organic chemicals whose concentrations in soils, water and living organisms have markedly increased in the second half of the twentieth century. As humans have striven to improve their living conditions, they have made increasingly large demands upon terrestrial and biological systems. Some of these effects have already been described; we shall now look at a number of cases which highlight the conflict between 'Humanity' and 'Nature'. Humans are inescapably a part of the natural system, and any disruption of 'Nature' inevitably affects ourselves.

Chemical pollution is understood to mean a high level of some chemical substance that adversely affects the natural environment. **Pollution** can be thought of as being the presence of too high a concentration of a resource in the wrong place at the wrong time. This emphasizes the view that pollution involves the misuse of resources: what may be correct for one system may be incorrect for a different system. The use of nitrate fertilizer is a good example of a chemical that may be either a useful resource or a pollutant. Nitrate added to the soil during the growing season is utilized by plants to give increased growth. The same nitrate fertilizer, if allowed to enter a water-supply reservoir, can make the water unsafe to drink, and lead to algal blooms and eutrophication in the reservoir.

There is a tendency to think of pollution as being only anthropogenic, but there are also natural events that produce harmful effects identical to those produced by human interference in the system. These natural events can also be classified as pollution. Natural oil seepages and high levels of toxic metals in soils due to the weathering of mineral deposits are two examples.

It is easy to recognize high-intensity pollution episodes such as the London smog of 1952 or the sinking in 1978 of the *Amoco Cadiz*, a large oil tanker, when 220 000 t of oil was released (a much larger quantity than the 35 000 t that escaped from the *Exxon Valdez* when it ran aground in Alaska in 1989). What are much more difficult to recognize are the slower changes and modifications to ecosystems that can occur as levels of potentially toxic chemicals build up. The problem may go unrecognized until it is either too late to reverse the effects or too late to save a substantial proportion of the affected population. This may be due to the natural variations in ecosystems making changes difficult to detect and to a long period of time elapsing between contamination occurring and the effect becoming noticeable. In the manufacturing, chemical and agricultural industries there have been numerous examples of substances being used that were later shown to be health hazards. One example is provided by the naturally occurring mineral asbestos: this has now been shown to be carcinogenic and its use has consequently been severely curtailed.

Asbestos

Asbestos exposure is associated with two types of cancer. Lung cancer is dose related: the greater the exposure, the higher the likelihood of lung cancer, and the lungs of these cancer sufferers have been found to contain about 1 g of asbestos. There is a **synergistic** relationship with smoking. One study of the life expectancy of workers exposed to asbestos has shown that, with the smokers, 50% had died of lung cancer before reaching the age of 53 years; whereas with non-smokers, 50% had died of lung cancer by the age of 72 years. This form of cancer can be reduced by dust suppression and the prevention of smoking.

The second type of cancer, called mesothelioma, is associated with much smaller exposures (accumulations of less than 1 mg being fatal) and is not related to smoking. What is even more worrying is that many sufferers were not employed in the asbestos industry, but were either members of the families of asbestos workers or lived in the vicinity of asbestos-using works. The period between exposure and recognition of the tumours varies between 15 and 50 years, with an average of about 40 years. Mesothelioma appears to be associated with amphibole asbestos fibres (Figure P4.1) such as crocidolite: these

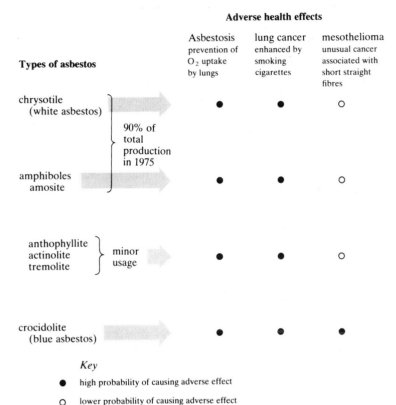

Figure P4.1 The different types of asbestos and their tendency to cause adverse health effects.

are straight and chemically more inert than chrysotile, which has a coiled configuration and which tends to disintegrate in the lungs as magnesium is leached out of it. Research into causes and effects has been slow due to problems in identifying the various types of asbestos and in relating the degree of exposure to the incidence of the disease, because of the long development time. Even if all use of asbestos was now banned the number of people dying from mesothelioma would be likely to go on rising due to exposure in the past 40 years.

Mesothelioma occurs in a village in Turkey where the local soft rock is used to build houses. In the period 1970–4, 70% of the deaths were due to mesothelioma and 10% were due to lung cancer. Though there is no asbestos in the dust, there are similar fibrous minerals of the correct size (0.1 μm diameter and 2–20 μm long) to cause cancer. It seems likely that all fibrous materials should therefore be viewed with concern.

Further problems highlighted by the asbestos case are that once inhalation has occurred there appears to be no way either of predicting whether an individual will actually develop mesothelioma or of preventing the cancer developing.

Though the hazards due to asbestos usage have taken a long time to become recognized, at least they are now well documented and steps can be taken to control the use of asbestos. Mesothelioma is so unusual that its increased occurrence was quickly noticed. If a pollutant produces an effect that is more subtle, such as reducing intelligence levels by 5% or increasing the incidence of bronchitis by 5%, there is a very much smaller chance that the change will be recognized. Thus the quality of life can be degraded without the effect or the cause being identified. Quality of life is a concept that extends to the lives and habitats of all living organisms, and gradual changes to these dynamic and often complex systems are equally difficult to detect.

Changes in cycling rates

The natural fluxes in all elements follow a cyclical path through the same major reservoirs. The differences between the sizes of the fluxes and the amounts in the reservoirs reflect both the chemical properties of the elements and the quantities of these elements actually present near the surface of the Earth. Human activities have increased a number of the fluxes associated with natural cycles, especially in the pathways that lead from the land-based to the ocean-based reservoirs.

Land use has been modified by building cities and transport systems, by agricultural developments and by deforestation. As a result erosion rates are now estimated to be two or three times what they were before humans changed from a mobile hunter–gatherer life-style to that of a settled community. Although it is difficult to be precise about the actual increase in rates of erosion, the loss of productive land has been estimated as $6 \times 10^4 \, \mathrm{km^2 \, a^{-1}}$ due to desertification, $7–20 \times 10^4 \, \mathrm{km^2 \, a^{-1}}$ due to deforestation, and $0.6 \times 10^4 \, \mathrm{km^2 \, a^{-1}}$ due to urbanization. These figures indicate that 0.1–0.2% of the ice-free land mass is losing its productive soil cover each year. These losses are seen to be very significant when one realizes that the world's population is growing at a rate of 1.7% per annum and that the *productive* land area is only about one-third of the *total* land area. The increased erosion rates increase the fluxes of all elements from land to ocean and also increase the loss of volatile elements (e.g. mercury and radon) from the land to the atmosphere.

Extraction of useful materials by mining, quarrying and subsequent processing has a much more specific and localized effect than the general increase in erosion from land-use change discussed above. The quarrying of building stone, the extraction of gravel and sand for building purposes, of shale and limestone for cement and chemicals, of phosphate rock for fertilizer, and of sodium and potassium chlorides for chemicals and fertilizers have between them mobilized much greater tonnages than has the mining of ores for metallic elements. However, we shall largely concentrate on the metals because they are more limited resources that are posing more acute environmental problems.

The mining of a deposit for one element may cause the release of other elements and compounds. For example, chalcopyrite, $CuFeS_2$, may be the economic mineral that is being mined as a source of copper. The concentration of the mineral in the ore is likely to be low, e.g. 2%, and only about one-third of the mineral is copper. This means that about 99.3% of the ore body is removed from the ground, broken up and then dumped. Elements other than the one being specifically mined are likely to be mobilized. In the chalcopyrite case, iron and sulphur are obvious examples, with the sulphur being converted to either sulphur dioxide or sulphuric acid.

Mining operations are transitory in nature. The lifetime of the mine is dependent upon the economic situation and the presence of the required ore. The term 'cut-off grade' is used to indicate the lowest concentration of the required element or mineral that can be economically exploited. The average cut-off grade varies widely from element to element – e.g. iron, 30%; lead, 2%; gold, 0.0003% – as do production rates, reserves and projected lifetimes of these reserves. The consumption of metals is much greater in the so-called developed countries than in the Third World countries. For instance, the USA, with 5% of the world's population, consumes 30% of the world's raw materials. If the underdeveloped areas become developed in the modern meaning of the term, consumption rates will rise rapidly and resources will be depleted that much more rapidly. Even with new processes, the energy required for the extraction of metals from less concentrated ores will rise very rapidly, creating resource pressures on all fuels, whether fossil, nuclear or biomass.

The extraction of ores followed by the reduction of the combined metal to its elemental form generally involves large inputs of energy and produces environmental problems with the disposal of the wastes. Both of these factors could be alleviated, and the resource-pressure problems could be reduced, by the recycling of the metals. As we have already seen with aluminium and steel, there are large energy savings to be made by recycling, but collection and separation of the required metals from the redundant manufactured articles can be very difficult. Quite large quantities of recycled scrap are already utilized, for instance, in Western Europe and the USA, about 50% of aluminium, 60% of copper, 40% of lead and 25% of steel are recycled.

Essential and toxic elements

Some elements have been found to be essential for life, but all elements are harmful at excessive concentrations. It must always be remembered that the chemical form of the element modifies its usefulness to an organism. Hydrogen, carbon and nitrogen are essential to humans, but they are useless in their elemental forms, and they are highly toxic if combined as hydrogen cyanide, HCN. It is therefore an oversimplification to state that certain elements are essential and other elements are toxic. The element must be in an available form for it to interact with an organism. Although elemental nitrogen comprises 80% of the atmosphere, it is not available to the majority of living organisms and so it is neither essential nor toxic: there is simply no interaction.

Nitrate, NO_3^-, is soluble and available to the majority of organisms. If plants are grown under conditions in which nitrate is the only source of available nitrogen, but in which all other nutrients are present in adequate amounts, then a growth–response curve of the type shown in Figure P4.2 will be obtained. In the region AC there is a deficiency of nitrate, and an increase in nitrate concentration gives an increase in yield. Over the concentration range CD there is a sufficient

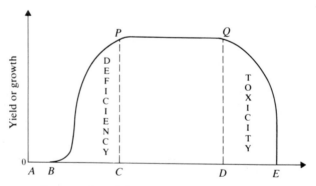

Concentration of nitrate, or any essential element,
available to plant

Figure P4.2   Idealized growth-response curve for the effect of changes in concentration of nitrate on plant growth.

supply of nitrate and yield is no longer limited by the nitrate concentration, hence the yield plateau *PQ*. After *Q* the yield begins to drop as the concentration of nitrate becomes so great that it interferes with other metabolic processes in the plant. At concentrations greater than *D*, the nitrate is toxic and above *E* the plants die: the nitrate has reached a lethal level.

If the experiment was repeated with humans, using nitrate as the only source of nitrogen, only the portion *AB* of the curve would be found – the nitrate nitrogen is for all practical purposes unavailable. If other sources of nitrogen were available, then the human response curve would resemble Figure P4.3. The initial plateau *ST* over the nitrate concentration range *LM* indicates that the body can control the amount of nitrate absorbed and keep it below a harmful level. As the concentration increases from *M* to *N* toxic effects would cause illness and eventually death; *N* is the lethal level.

Curves similar to that shown in Figure P4.2 are obtained with all essential elements when present in available forms. In fact, essentiality is *recognized* by the production of this type of curve. Non-essential elements and elements in unavailable forms would produce curves similar to Figure P4.3. The greater the toxicity of the chemical, the shorter the stable plateau *ST*.

The actual expression of the toxicity of chemical species is very difficult, especially with regard to humans. Toxicity is generally expressed in terms of the dose that kills 50% of a population, i.e. the LD_{50} value. ('LD' means 'lethal dose'.) Provided a large enough population is studied, individual variability in response, which can be very great, is taken into account. However, toxicity effects are

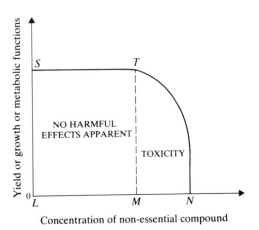

Figure P4.3 Idealized growth-response curve for non-essential and toxic elements.

influenced by the mode of application (e.g. whether by mouth, lungs, skin, or injection) and by the presence or absence of other compounds or foodstuffs. A dose of 0.02 mmol $HgCl_2$ kg^{-1}, which killed 39 out of 40 rats, caused only 1 death out of 40 rats that were also given sodium selenite, Na_2SeO_3. LD_{50} values are at best only indicators of relative toxicity. Obviously toxicity testing cannot be carried out on humans, so there is very little detailed knowledge of toxic doses for humans. Information gained from studies using other animals such as rats or rabbits is notoriously difficult to interpret in relation to humans. For instance, the rate of gastrointestinal absorption of lead ($Pb^{2+}_{(aq)}$) by rats is only about one-tenth of that found in humans.

Further controversy has been generated over the effects of low levels of potentially toxic chemical compounds. There is a lot of evidence that elements such as lead, mercury and cadmium accumulate with time in some organisms. However, it is not clear at what stage point T (Figure P4.3) is reached and adverse effects are felt. The increased mobilization of elements plus the production of compounds not usually found in natural systems put increased strain on the ability of living organisms of all types to avoid slipping off the ST plateau.

Despite all the publicity given to chemical pollution – local *excesses* of elements – with both its actual and possible effects, there is no doubt that *deficiencies* of both major and trace essential elements are a much more widespread cause of sub-optimal growth. The essential elements are required to make up the chemical compounds that together constitute a living organism. There are 11 major essential elements (hydrogen, carbon, nitrogen, oxygen, sodium, magnesium, phosphorus, sulphur, chlorine, potassium and calcium), and there are about another 15 elements (boron, fluorine, silicon, vanadium, chromium, manganese, iron, cobalt, nickel, copper, zinc, selenium, molybdenum, tin and iodine) that are ordinarily present in plants or animals in concentrations of less than 0.01% but which are essential to some or all organisms – e.g. boron appears to be essential to plants but not to animals.

Consider the distribution of the trace elements in soils (Figure P4.4). Some forms are more available for uptake by plants than others. The soluble ions and complex ions are directly available providing the plant root has no specific exclusion mechanisms. The adsorbed ions are likely to be the next most available form as in many cases they are readily displaced from their sorption sites, particularly if conditions become more acid and hydrogen ions displace these cations. The availability of the trace elements dispersed in the hydrated oxide precipitates is increased by more reducing conditions (e.g. on water-

Solid

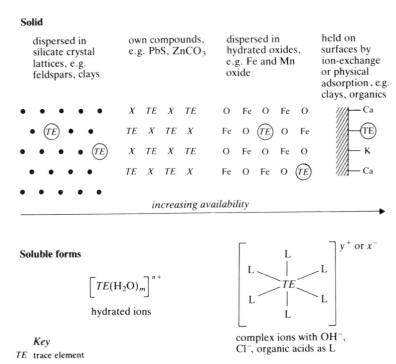

| dispersed in silicate crystal lattices, e.g. feldspars, clays | own compounds, e.g. PbS, $ZnCO_3$ | dispersed in hydrated oxides, e.g. Fe and Mn oxide | held on surfaces by ion-exchange or physical adsorption, e.g. clays, organics |

increasing availability

Soluble forms

$$\left[TE(H_2O)_m \right]^{n+}$$

hydrated ions

$$\left[\begin{array}{c} L \\ L \diagdown \big| \diagup L \\ TE \\ L \diagup \big| \diagdown L \\ L \end{array} \right]^{y^+ \text{ or } x^-}$$

Key

TE trace element

complex ions with OH^-, Cl^-, organic acids as L

Figure P4.4 The major forms in which trace elements are found when distributed in soils.

logging) or by other changes, such as increased acidity, that also increase the solubility of the hydrated oxide. Availability of the trace elements from their own insoluble compounds, or when dispersed in other minerals, depends upon the relative stability of the compound or mineral.

It is clear that because a high proportion of a particular element may be bound up in the minerals of the soil in an unavailable form, analysis of the total element concentration does not necessarily provide a useful indication of whether plants growing on that soil will suffer from either deficiency or toxicity problems. How to determine the available levels of elements in soils and sediments is one of the major problems of analytical chemistry. The most common approach to this problem is to use different extracting solutions that are supposed preferentially to extract particular chemical forms from the soil. For example, ammonium acetate solutions are used to indicate easily exchangeable metals: the NH_4^+ ions will occupy the cation-exchangeable sites (Chapter 12), releasing the metals previously bound to these sites. Similarly, solutions of the complexing agent EDTA will remove more tightly held ions associated with organic matter. The use of a series of these 'selective' extracting agents gives a rough indication of how the metals are bound in soils and sediments.

13 Lead

Lead Abundance by weight (the relative abundance is given in parentheses): crust, 13 ppm (35); ocean, 0.03 ppb (46).

As can be seen from the geochemical cycle of lead (Figure 13.1), the rate of extraction is about ten times greater than the natural rate of weathering. The two major uses of lead are in lead–acid storage batteries, particularly for motor vehicles, and as lead alkyl compounds added to petrol. Petrol engines give higher-power outputs and use fuel more efficiently when operated at higher compression ratios. Smooth running can only be achieved under these conditions by using petrol containing more branched-chain or aromatic hydrocarbons or by the cheaper expedient of adding lead alkyl compounds. Catalytic convertors installed in motor vehicle exhaust systems to cut down emissions of nitrogen oxides, hydrocarbons and carbon monoxide require lead-free petrol to be used. If leaded petrol is used, the lead compounds in the exhaust emissions are trapped on the catalyst surface and prevent the catalyst reacting with the other components in the exhaust gases. Environmental legislation and taxation policy have resulted in an increase in sales of lead-free petrol in the UK and Europe and the amount of lead being released into the atmosphere is declining. The majority of the lead used in batteries is recycled and only causes problems when the battery-disposal and recycling processes are not effectively controlled. In contrast, about 75% of the lead added to petrol is emitted through the exhaust and dispersed as an aerosol in the atmosphere. As a consequence it is unlikely that there is anywhere left on the Earth's surface that has 'natural' levels of lead. Results from Greenland ice cores indicate that there has been a 400-fold increase in lead deposition in the ice between 800 BC and 1965. The average anthropogenic emission rate in the latter half of the nineteenth century was 22×10^6 kg a^{-1}, due mainly to the smelting of lead ores and burning of coal. Nowadays the rate is about 20 times higher, at 450×10^6 kg a^{-1}. Approximately 94% of the lead in the atmosphere is derived from anthropogenic sources, with an even higher proportion in urban areas where there is heavy motor traffic.

Though lead ores may be quite rich in lead, the cut-off grade is

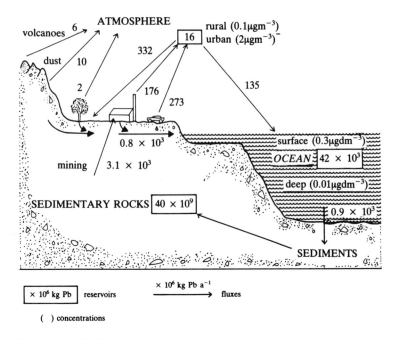

Figure 13.1   The lead cycle.

about 2%; the average concentration in rocks (13 mg kg^{-1}) and soils (20 mg kg^{-1}) is low. The enrichment factor needed to produce a lead ore is about 1500:

$$\text{enrichment factor} = \frac{\text{concentration of lead in ore}}{\text{concentration in crust}} = \frac{20 \times 10^{-3}\,\text{kg kg}^{-1}}{13 \times 10^{-6}\,\text{kg kg}^{-1}} \simeq 1500$$

When this figure is compared with the enrichment factors for iron and aluminium, about 5 and 4 respectively, it is apparent that unusual conditions must be required for lead-ore formation – hence the concern about the size of future reserves.

As well as being in vein ore deposits, lead sulphide is also associated with black shales. These fine-grained marine deposits formed under reducing conditions in which sulphate was reduced to sulphide (Eqn 13.1) and the insoluble sulphides of metals like lead, copper and mercury were precipitated (Eqn 13.2).

$$SO_{4(aq)}^{2-} + 8e^- + 8H^+ \longrightarrow S_{(aq)}^{2-} + 4H_2O \qquad (13.1)$$

$$Pb_{(aq)}^{2+} + S_{(aq)}^{2-} \longrightarrow PbS_{(s)} \qquad (13.2)$$

Lead sulphide has a low solubility ($10^{-27}\,\text{mol dm}^{-3}$) and the solubility of other lead compounds is usually less than $10^{-6}\,\mathrm{mol\,dm^{-3}}$. The weathering of lead ores generally produces immobile compounds, with low concentrations of lead in natural waters. The majority of lead in rocks is present in silicate structures, substituting for elements like calcium; it is unavailable to plants until the primary mineral is broken down by weathering. In soils the lead tends to be held strongly by organic matter. It is often concentrated in the upper few centimetres due to incorporation in plants and then accumulation within the decomposing remains in the humus layer. It seems likely that some reports of surface contamination by atmospheric lead causing much higher levels in the top layers of soil were actually misinterpretations of the natural biogeochemical concentrating effects. About 95% of the lead being transferred to the oceans by rivers was transported as suspended sediment (inorganic and organic). The increase in aerial dispersion from anthropogenic emissions has increased the atmospheric flux, leading to higher levels of lead in rivers and some lead falling directly on the sea.

Lead was first used to produce water pipes in Roman times, and until the 1950s was still used extensively in the UK. Since then lead pipes have been largely replaced by copper ones, which are easier to fabricate and to bend; and, more recently, in cold-water systems by plastic ones, plastic being cheaper than copper. In hard-water areas, in which the pH of the water does not drop below 7, lead pipes are quite stable. A layer of lead, calcium and magnesium carbonates is rapidly built up and acts as a protective coating against dissolution of the lead. In soft-water areas, in which the pH of the water may be below 5, the lead is relatively soluble: water that has been in contact with the lead for a period of time may contain over $1\,\mathrm{mg\,Pb\,dm^3}$. The EC limit for lead in drinking water is $50\,\mu\mathrm{g\,dm^{-3}}$, but the WHO is now recommending that the limit should be reduced to $10\,\mu\mathrm{g\,dm^{-3}}$. This will be very difficult to achieve if there is any lead piping in the supply system. Water authorities in the UK now increase the pH of all acid waters by the addition of calcium hydroxide, $Ca(OH)_2$. This also increases the hardness of the water. Lead can cause hypertension, one of the factors influencing heart disease, and it has been suggested that the lower levels of heart disease in hard-water areas may be due to lower levels of lead in the water supply. The importance of avoiding acid water in lead pipes was illustrated during the 1983 strike of the UK water workers. The strike prevented the treatment of many water supplies with calcium hydroxide. Within 2 weeks the levels of lead in the drinking water of some houses with extensive lead piping had risen from $30–40\,\mu\mathrm{g\,dm^{-3}}$ to $800–1200\,\mu\mathrm{g\,dm^{-3}}$. Even houses with copper pipes are not completely free from contamination by lead because the soldered joints between the pipes contain lead that can be mobilized by acid waters.

Dichloroethane, $C_2H_4Cl_2$, and dibromoethane, $C_2H_4Br_2$, are added to petrol to prevent the build-up of lead deposits in petrol engines running on petrol containing lead-alkyl compounds. The lead combines with the chlorine or bromine to form volatile halogenated compounds that are removed from the engine with the exhaust gases. The cooled compounds form fine particles (mainly less than $2\,\mu m$ diameter) which can be widely dispersed and which can easily be breathed into the lungs.

Lead is a cumulative poison. About 90% of the lead retained in the body enters the bones, from which it can be remobilized. The WHO recommends a maximum intake of 3 mg lead per person per week. Children and infants, who are more susceptible than adults to lead poisoning, should have intakes of less than 1 mg Pb per week. The average rate of absorption of dietary lead is about 8%, but about 40% of the fine particulate lead retained in the lungs is absorbed (Figure 13.2). These absorption rates result in about two-thirds of absorbed lead coming from the diet and one-third from the atmosphere. In addition lead intake is increased by about 5% for every 20 cigarettes smoked per day. The absorbed lead enters the bloodstream, where over 90% is bound to the red blood cells with a mean residence time of 1 month. The blood-lead level appears to vary linearly with exposure and is therefore a good indicator of recent exposure, with normal levels being $10-20\,\mu g$ Pb per $100\,cm^3$ blood; 25–40% of the lead

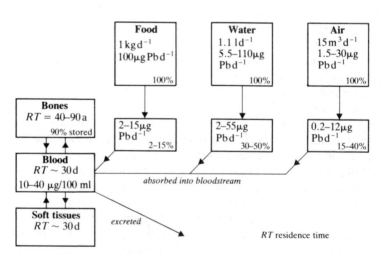

Figure 13.2   The daily intake of lead by adult humans, and its distribution in the body.

in blood enters soft tissues, about 15% enters bones, and the remainder is excreted. The residence time in the soft tissues is again about 1 month, with the majority of the lead being lost in hair, perspiration and digestive secretions. The concentration in hair is a good indicator of exposure to lead over the previous few months. Bones act as the major reservoir for lead in the body and the residence time is 40–90 years in adults. Children absorb a greater proportion of ingested lead than do adults, and the distribution in the body is also different. About 28% of the lead in children is in soft tissues compared with only 5% of the lead in adults being in the soft tissues.

Lead binds strongly to a large number of molecules, such as amino acids, haemoglobin, many enzymes, RNA and DNA; it thus disrupts many metabolic pathways. The effects of lead toxicity are very wide ranging and include impaired blood synthesis, hypertension, hyperactivity and brain damage. Effects at the molecular level, such as the interference with steps in haem synthesis, occur at very low levels equivalent to $20–200\,\mu g\,Pb\,kg^{-1}$ soft tissue, which are the levels normally found in adults. If the normal body burden in adults is of the same order as that at which harmful effects may occur, any factor that increases lead intake must be viewed with great concern. The widespread distribution of lead from motor-vehicle exhausts increases atmospheric levels by factors of 20 (much more in urban areas); further, the subsequent contamination of soil and crops increases the amount of lead in food. Children, as well as absorbing a higher proportion of ingested lead than adults, have poorer hygiene habits and consume more dirt and dust picked up on their hands. In urban areas roadside dirt can contain several thousand parts per million lead (typical mean values $1000–3000\,\mu g\,kg^{-1}$) and samples of old paint in West London contained from 6.2% to 52.4% lead. Some children eat dirt, a habit known as pica, and they are particularly at risk from lead paints, one piece of which may contain enough lead to cause acute poisoning. Acute poisoning produces a rapid onset of symptoms due to the uptake of a large dose of toxic material. Chronic poisoning is due to the uptake of smaller doses over a longer period of time and generally brings about a gradual appearance of symptoms. Associated with chronic poisoning are sub-clinical effects in which there are no obvious outer (or clinical) signs of adverse effects, but metabolic functions at the molecular level are being disrupted to a greater or lesser extent. Because of the lack of obvious symptoms, sub-clinical effects are difficult to diagnose. There is a growing body of evidence that the intelligence of children is being lowered in urban areas in which there are high inputs of lead from motor vehicles.

Blood-lead levels are commonly used to monitor the status of lead body burdens. The results are sometimes not very satisfactory because there is not always a close relationship between blood-lead levels, soft-tissue levels or concentrations that cause metabolic dysfunction. One individual with $25\,\mu g$ Pb per $100\,cm^3$ blood may be suffering from lead-toxicity effects, whereas another with $70\,\mu g$ Pb per $100\,cm^3$ blood may not. As it is individuals, rather than the statistical average members of the populations, that must be treated, a monitoring programme should ideally pick out the individuals at risk – otherwise either individual sufferers will be missed or resources will be wasted on the treatment of too many unaffected people. However, the use of blood sampling will continue as there is no other sampling technique that yields better results and is so easy to carry out.

The analysis of low levels of lead is difficult. In natural waters lead levels are at the microgram per cubic decimetre level. There are problems (a) in achieving the high degree of sensitivity required actually to detect the lead, (b) in avoiding contamination from lead in the sampling containers, in the filter units, in the chemicals used and in the laboratory atmosphere, and (c) in overcoming interference from other species present in the solution (this is especially a problem with sea water). In one inter-laboratory trial, all seven laboratories taking part reported sea-water lead levels 10–100 times higher than the correct values. In solid samples, from which the lead has to be extracted into a solution, the errors can be just as great even though the original concentrations may be somewhat higher. One of the commonly used analytical technique for lead is prone to give readings that are too high unless great care is taken in providing careful corrections. It appears likely that many reported analyses for lead are incorrect. However, there is now a much greater emphasis on quality control and the use of reference samples so that published results are generally more reliable than they were before the mid-1980s.

14 Mercury

Mercury Abundance by weight (the relative abundance is given in parentheses): crust, 80 ppb (63); ocean, 0.15 ppb (40).

Mercury is much less common than lead in the Earth's crust and consequently the geochemical cycle has smaller fluxes (Figure 14.1). Mercury is the only metal that is liquid at normal temperatures, being 15.5 times more dense than water. It has a high volatility: air that is in equilibrium with liquid mercury will contain $14 \, \text{mg Hg m}^{-3}$ air at $20 \, ^{\circ}\text{C}$. The maximum allowable level of any toxic vapour – the level that is thought to be safe – is called the **threshold limit value (TLV)**. For mercury, this concentration is set at $0.05 \, \text{mg Hg per m}^3$ air. All mercury spills are potentially very dangerous.

Because of mercury's association with lead–zinc–silver ores, the presence of high levels of mercury vapour in the atmosphere, or in the air trapped in soil, can be used to indicate the presence of these ore bodies. Mercury-vapour concentrations can be determined by

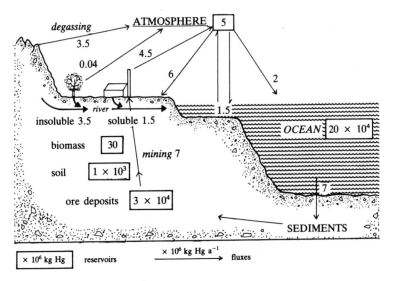

Figure 14.1 The mercury cycle.

very sensitive portable instruments which allow the method to be used extensively in mineral exploration. Mercury is described as a 'pathfinder' element when used in this way to indicate Pb–Zn–Ag mineralization. Most mercury compounds are relatively volatile; even mercuric sulphide, HgS, which is the major mercury ore mineral, cinnabar, gives 10 ng Hg m^{-3} in dry air.

Cinnabar is easily reduced to the metal by heating in air (Eqns 14.1 and 14.2) and as elemental mercury is also found in small amounts, the metal has been used by humans for several thousand years.

$$2HgS + 3O_2 \xrightarrow{\text{heat}} 2HgO + 2SO_2 \qquad (14.1)$$

$$2HgO \xrightarrow{500\,^\circ C} 2Hg + O_2 \qquad (14.2)$$

The production of mercury has declined since the early 1970s. In 1971 it was 9.6×10^3 tonnes. By 1989 it had dropped to 5.4×10^3 tonnes. This was partly due to increased recycling and partly to the concern about environmental pollution, which led to decreased usage. The largest single use of mercury in 1971 was in the manufacture of sodium hydroxide and chlorine by the electrolysis of brine. The losses of mercury from these plants were 150–250 g Hg t^{-1} Cl$_2$ produced. Strict emission controls have since been introduced and these have resulted either in alternative processes for chlor-alkali production or in the reduction of losses and the amounts of replacement mercury used.

The most important difference between the lead and mercury cycles is the importance of methylation reactions. In its natural cycle lead remains in the +2 state and little natural methylation occurs. Mercury exists in the 0, +1 and +2 oxidation states (Figure 14.2), and methylation (Eqn 14.3) is an important feature of its cycle, particularly with regard to its uptake by fish and humans.

$$Hg^{2+} \xrightarrow{\text{micro-organisms}} \underset{\substack{\text{monomethylmercury} \\ \text{soluble in water}}}{CH_3Hg^+} \xrightarrow{\text{micro-organisms}} \underset{\substack{\text{dimethylmercury} \\ \text{insoluble in water} \\ \text{volatile}}}{(CH_3)_2Hg} \qquad (14.3)$$

Methylmercury is the major mercury species found in fish and about 90% of the CH$_3$Hg$^+$ eaten is absorbed by humans. There are a variety of mechanisms by which micro-organisms can bring about the formation of the very toxic methylmercury species (Figure 14.2). Transformation to monomethylmercury is more important under aerobic conditions than under anaerobic conditions. This is because the presence of sulphide ions (S^{2-}) in reducing environments favours the production of insoluble mercury(II) sulphide. Rates of methylation of insoluble mercury species are very much slower than with

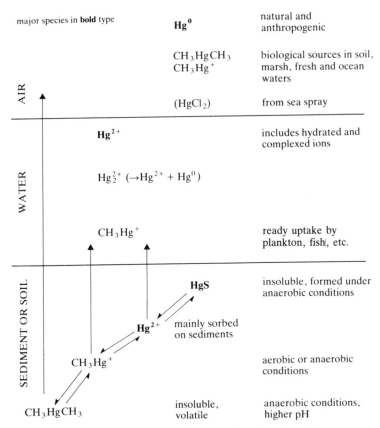

Figure 14.2 Mercury species found in air, water and sediments or soil.

soluble species. The methylation of mercury is especially important in rivers and lakes that have a low pH either continuously or intermittently. This is because:

(a) more Hg^{2+} is solubilized by ion exchange from sediments at high hydrogen-ion concentrations and the rate of synthesis of methylmercury by micro-organisms is therefore enhanced;

(b) the low pH favours methylmercury synthesis rather than dimethylmercury synthesis;

(c) the lower population of higher organisms means that each individual absorbs a greater proportion of the methylmercury produced.

In very productive water bodies (that have a higher pH) there is a much greater dilution of the methylmercury as it is spread through a much larger number of individuals.

As well as the micro-organisms that *produce* methylmercury there are many micro-organisms that *break down* methylmercury, so normal levels do not rise very high in water bodies. The concentration of methylmercury in the water reflects the balance between methylation and demethylation reactions. The concentration of methylmercury in aquatic organisms varies from species to species but appears to be related to metabolic activity and average lifespan. Methylmercury makes up 60–90% of the mercury in fish because the degree of absorption is high and the rate of elimination is usually very low. Under similar conditions only 10–15% of the inorganic mercury is absorbed and elimination of $Hg^{2+}_{(aq)}$ occurs freely.

There have been a number of tragic episodes of organo-mercury poisonings in Japan, Guatemala, Iraq and Pakistan; these awakened interest in mercury as an environmental pollutant. In each case the causative factors were the consumption of food contaminated by industrially produced organo-mercury compounds. For instance, over 3000 people died in Iraq in 1971–2 as a result of eating seed grain treated with methylmercury fungicides. Investigations resulting from these cases led to the realization that methylmercury compounds could be produced in natural systems. This was an important discovery: it had been thought previously that the majority of mercury released by human activities would be either sorbed on to inorganic solids or converted into insoluble mercury(II) sulphide, or both. These mechanisms would effectively remove mercury from interactions with ecosystems. Biomethylation means that the mercury, rather than being locked up indefinitely, can be remobilized and is available for uptake by aquatic organisms. At present the intake of mercury from fish is the only major source of human dietary mercury, but the use of fishmeal for feeding livestock and poultry could result in high levels of methylmercury in milk, meat and eggs.

The target organ of methylmercury in humans is the brain, where it disrupts the blood–brain barrier, upsetting the metabolism of the nervous system. In addition, mercury forms very strong bonds with sulphur groups in proteins and enzymes, disrupting various enzymatic systems and synthetic mechanisms. The main toxic effects of inorganic mercury are that it tends to disrupt the functions of the kidneys and liver. Compared with inorganic mercury, methylmercury can much more easily cross the placenta and affect the foetus. In the poisoning case in Minamata, Japan, in which a factory released methylmercury into an estuary and so contaminated fish, 22 infants with serious brain damage were born to mothers who showed slight or no symptoms of mercury poisoning.

The WHO has suggested that the intake for adults should be less than 0.3 mg total mercury per person per week, of which no more than 0.2 mg should be methylmercury. Drinking water normally contains well below the guideline (1 μg Hg dm^{-3}), and plants do not appear to take up much mercury from the soil. The average intake is about 0.07 mg Hg per person per week. However, people who eat a lot of fish may consume much more – for instance, levels of 0.6 mg Hg kg^{-1} fish could provide 0.15 mg methylmercury in one meal.

Despite the increased dispersion of mercury by mining, use of mercury compounds and fossil-fuel combustion, there is little evidence of worldwide contamination on the same scale as has occurred with lead. For instance, polar ice cores that showed lead levels increasing from 10 ng kg^{-1} in 1750 to about 200 ng kg^{-1} in 1960 indicated mercury levels of about 100 ng kg^{-1} since 800 BC. The difference between the two metals can be explained by the much larger natural atmospheric component of mercury due to terrestrial degassing, masking the effects of the anthropogenic inputs. With lead, a much greater tonnage is put into the atmosphere by anthropogenic sources than by natural sources. The evidence that mercury is actually causing widespread health problems is very sparse. The people who are definitely at risk are those working in mercury-using industries, where health and safety regulations should be enforced to protect them. Fish eaters may be at risk, but even here there are many people who have consumed fish containing 0.5–1.0 mg Hg kg^{-1} who do not appear to be adversely affected. The known chronic or acute toxicity cases in Iraq, Guatemala and Pakistan have involved seed grain treated with organo-mercury fungicides that has been eaten rather than planted, and the Japanese cases produced methylmercury contents of 20 mg Hg kg^{-1} in the fish. The optimistic view is that stricter environmental standards and future vigilance will prevent the widespread problems that are more likely with lead and possibly cadmium.

There are very sensitive methods available that will detect mercury at levels in the microgram per kilogram range, but the high volatility of many mercury species causes difficulties. Samples must be collected in a manner that will avoid contamination, and the sample must be representative of the mass being investigated. The sample must be stored and then converted into a form suitable for analysis. Each stage can cause problems. During storage mercury can be lost from the sample by diffusion of gaseous forms into the air above the sample or out of the container, or there may be sorption of species on to the container walls. An aqueous solution stored as collected for 21 days lost 77% of the original mercury by sorption on to the container walls and a further 18% by volatilization. Another sample of the same solution to which potassium dichromate and nitric acid were added only lost 1% by sorption and 1% by volatilization in the same time. The potassium dichromate oxidizes volatile species to mercury (II) ions and the hydrogen ions from the acid compete for the sorption sites on the container walls. Solid samples usually have to be dissolved before the analysis can be carried out and losses of mercury can easily occur, especially if the process requires heating. Closed containers or very efficient condensers can overcome this problem.

15 Zinc and cadmium

Zinc Abundance by weight (the relative abundance is given in parentheses): crust, 70 ppm (24); ocean, 11 ppb (22).

Cadmium Abundance by weight (the relative abundance is given in parentheses): crust, 0.2 ppm (64); ocean, 0.11 ppb (42).

Zinc is an essential element, but cadmium has no known useful biological function in humans. The cycles of zinc and cadmium (Figure 15.1) are very closely inter-related because natural zinc minerals and most anthropogenic fluxes contain small amounts of cadmium. Zinc, cadmium and mercury are in the same group in the periodic table because they all have similar arrangements of electrons in the outermost shells. However, the inner electron structure of mercury differs from that of zinc and cadmium, and therefore its chemical properties differ also; the properties of zinc and cadmium, though, are very similar. The hydrated zinc ion ($Zn^{2+}_{(aq)}$) is relatively

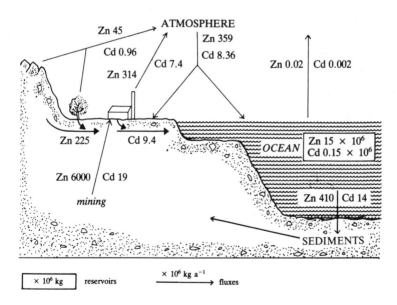

Figure 15.1 The zinc and cadmium cycles.

more stable than the hydrated cadmium ion ($Cd^{2+}_{(aq)}$). As a consequence, when there is competition between zinc and cadmium to bind to sites in insoluble species, cadmium tends to bind more strongly than zinc, especially if a metal–sulphur bond is formed. Cadmium therefore displaces zinc from many of the zinc-containing enzymes.

The mobilization of zinc and cadmium by human activities outweighs natural cycling processes. The increased quantities of cadmium mobilized are partly related to demand for cadmium itself but also to increased use of zinc and phosphate fertilizers. Cadmium is used in electroplating (in which it forms a bright corrosion-resistant finish), in plastic stabilizers, in pigments, in solder, and in nickel–cadmium batteries. Zinc is used to prevent corrosion (by galvanizing), in alloys, in paints, in dyes and in tyres. Cadmium is obtained as a byproduct from zinc refining (Eqn 15.1) but the complete removal of all the cadmium from 'bulk' zinc is uneconomical.

$$ZnS \xrightarrow[\text{heat}]{O_2} SO_2 + ZnO \xrightarrow[\text{heat}]{C} CO_2 + Zn/Cd \xrightarrow{\text{distil}} Zn + Cd \quad (15.1)$$

Dispersal of zinc therefore leads to dispersal of cadmium.

Phosphate fertilizer contains $5–100\,mg\,Cd\,kg^{-1}$ and this increases soil and plant levels when these fertilizers are used. It has been estimated that in the UK the average addition of cadmium to crop lands is $0.4\,mg\,m^{-2}\,a^{-1}$ (air deposition, 41%; phosphate fertilizers, 54%; sewage sludge, 5%). However, when sewage sludge is used over 90% of the added cadmium can come from the sludge. The lower average figure is because sewage sludge is only applied to about 5% of the land in the UK. The high values in sewage sludge reflect industrial and domestic usage of cadmium.

15.1 Sewage-sludge disposal

The problems of sewage-sludge disposal are becoming increasingly great for a number of reasons. As with most environmental problems, increased populations and the desire for higher standards of living have converted a relatively small-scale problem into a major resource conflict. This is summarized below.

Problem   How to dispose of human excreta.
Solution   Collect the wastes and spread them on land.

Problems   Pathogenic bacteria, viruses, parasites cause endemic diseases. Pollution of watercourses causes diseases. Larger population centres make collection and disposal more difficult. The smell of the collected material may be offensive.

Solution   Build a system of water closets and sewers.

Problems   The untreated sewer effluent uses up the dissolved oxygen in rivers, killing all higher aquatic life-forms. The smell and visible appearance of rivers deteriorates. The contamination of water supplies of downriver communities occurs.

Solution   Build sewage treatment plants (Figure 15.2).

Problems   The cost increases very rapidly as the type of treatment is uprated from primary to secondary to tertiary. The final water quality depends upon the degree of treatment. The solids still have to be disposed of.

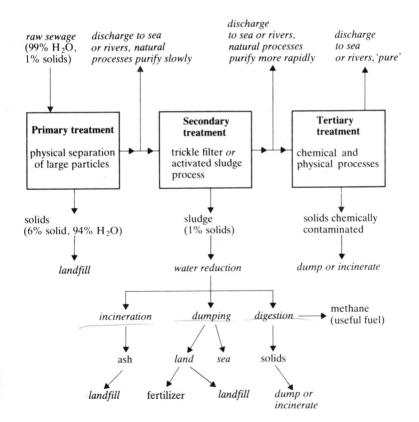

Figure 15.2   Flow diagram for sewage treatment.

Table 15.1 Comparison of metal concentrations in dry sewage sludge and soil (concentrations in mg kg^{-1})

Concentration in sewage sludge			Mean soil concentration	Mean sludge concentration / Mean soil concentration
metal	mean	range		
Hg	2	0–5	0.25	8
Cd	20	0–300	0.4	50
Cu	250	1–3 000	50	5
Cr	500	10–5 000	50	10
Pb	700	50–5 000	25	28
Zn	3 000	500–25 000	100	30
Fe	16 000	2 000–42 000	30 000	0.5

Solution The solids can be dumped on land, as they contain useful nitrogen, phosphorus, potassium and organic matter, to replace fertilizers.

Problems Urban and industrial areas produce sludge with high toxic-metal contents (Table 15.1). Therefore only limited amounts can be applied without damaging crops or consumers, and they may still contain pathogenic organisms.

Solution The solids could be dumped in the sea away from coast.

Problem The toxic contents will affect marine organisms.

Solution The sludge could be incinerated.

Problems This is costly because a lot of energy is needed to remove water before the sludge will burn. Air pollution may occur if gases not properly cleaned. Ash will contain even higher concentrations of toxic metals and must be dumped in special sites to prevent pollution of watercourses.

Final solution? Reduce the population; reduce the industrial contamination of the effluent; use smaller quantities of potentially toxic compounds.

Interim solution? Maximize the use of natural systems to detoxify contaminants and remove other toxic substances by treatment, preferably at source where their concentration is highest and other impurities lowest.

It should be noted that metals such as iron, zinc and copper are essential elements but they become toxic if present in too high a concentration. The effects of dumping sludge on land are still not clearly understood. Uptake of the metals by plants is dependent upon

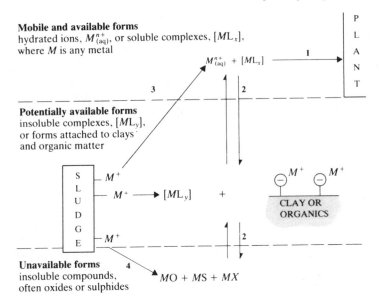

Mobile and available forms
hydrated ions, $M^{n+}_{(aq)}$, or soluble complexes, $[ML_x]$,
where M is any metal

Potentially available forms
insoluble complexes, $[ML_y]$,
or forms attached to clays
and organic matter

Unavailable forms
insoluble compounds,
often oxides or sulphides

Key

1   uptake by plant

2   changes in form due to action of
    micro-organisms or changes in pH or Eh

3   soluble forms pass directly into solution

4   formation of insoluble compounds

Figure 15.3   The availability of metals from sewage sludges.

their availability and mobility (Figure 15.3). Soils with a high pH and a high cation-exchange capacity will tend to immobilize most added metals. However, a later change in conditions, particularly a drop in pH, could cause the release of the metals. It is important to limit the build-up of potentially toxic elements even if they are initially found in unavailable forms. There are now limits recommended by the EC for the maximum concentration of some metals in soils that have had sewage sludge added to them. For soils of pH 6–7, the allowed maximum concentration for cadmium is 3 mg kg^{-1} and for zinc and lead 300 mg kg^{-1} each. In areas with naturally high background concentrations of these elements very little sewage sludge can be disposed of by dispersal on land. Because cadmium is chemically similar to zinc, it is also readily taken up by plants and enters the edible portions. This is an important difference from lead, which is not easily accumulated by plants. The concentrations of lead in the edible portions of plants are much lower than soil-lead levels. Cadmium concentrations in plants are much closer to soil levels, though this varies very much from species to species, and the relative uptake of cadmium is reduced as the concentration of cadmium in the soil increases. The soil: grass cadmium ratio changed from 3.3 to 11.4 to 157.1 as the soil-cadmium levels changed from 3 to 24 to 440 mg Cd kg^{-1} soil in a study carried out at Shipham, in Somerset.

15.2 Toxicity of cadmium

The toxic effects of cadmium received widespread attention as a result of some Japanese developing Itai-Itai byo ('ouch ouch' disease). The name came from the severe pain developed by the sufferers as lumbago-type pains progressed to become severe bone damage with multiple fractures of the softened bones. Death was attributed to kidney failure. The victims were mainly post-menopausal women suffering from malnutrition, low vitamin D intake and calcium deficiency. These contributing factors mean that Itai-Itai disease is not typical of cadmium toxicity. The main target organs for cadmium are the kidney and liver, with critical effects occurring when a content of 200 μg Cd g^{-1} (wet weight) is reached in the kidney cortex. The body burden of cadmium increases with age, as the half-life in the body is about 20–30 years. Normal dietary intakes are in the range 0.21–0.42 mg per week which are close to the WHO recommended maximum (0.4–0.5 mg per week). The closeness between actual intakes and suggested maxima is one of the reasons for the concern over cadmium levels in soil, water and food. Smokers are especially at risk because of the cadmium content of tobacco. The smoking of 20 cigarettes per day corresponds to an oral intake of 40 μg Cd from food, i.e. the cadmium uptake is doubled.

The problems in Japan arose from a lead-zinc mining and smelting operation upstream from where the victims lived. There was aerial pollution from cadmium-containing fumes and particulates, plus water pollution due to cadmium-containing sediments and mine wastes. Rice grown in paddy fields flooded by the river contained up to 3.4 mg Cd kg^{-1} rice. Because of its lower boiling point, cadmium is more concentrated, relative to zinc, in the atmospheric emission from zinc smelters than it is in the ore or in zinc metal. In the UK the position of zinc smelters is clearly marked in geochemical maps as cadmium anomalies.

15.3 Zinc

Zinc is an essential component of about a hundred enzymes in total though the number of zinc-containing enzymes in vertebrates is much smaller. Normal zinc levels in plants are $25–150\,\mathrm{mg\,kg^{-1}}$ with concentrations above $400\,\mathrm{mg\,kg^{-1}}$ being toxic. Human adults contain 1.4 g to 2.3 g zinc, about half the amount of iron. Zinc deficiency in humans leads to dwarfism, reduced rates of blood clotting and wound healing, skin abnormalities, and other problems.

Both zinc and cadmium are found associated with a soluble low-molecular-weight protein called metallothionen which is characterized by a high cysteine content. The large number of sulphydryl groups, $-SH$, bind the heavy metals tightly. One action of metallothionen is to detoxify cadmium, mercury and lead by binding very strongly to these metals through metal–sulphur bonds so that they are no longer available to interfere with other metabolic processes. Unfortunately the amount of metallothionen available is too limited to prevent the toxic effects of large intakes of the heavy metals.

16 Radon

Radon Abundance by weight: crust, 4×10^{-13} mg kg^{-1}.

Radon is responsible for about half of the total radiation dose of 2.5 mSv (milli**Sievert**) that it has been estimated that the average person in the UK receives (Table 16.1). It has also been estimated that exposure to radon is responsible for about 2500 UK deaths from lung cancer each year. Radon is a member of the inert, or noble, gas series (helium, argon, neon, xenon, krypton and radon). It has a very low chemical reactivity and under normal natural conditions it is an unreactive, odourless, colourless gas (boiling point 211 K, -62 °C), with a density almost eight times greater than that of air.

Radon is one of the elements that have no known stable isotopes. The stability of the nucleus is dependent upon the attractive forces between the nucleons (neutrons and protons) outweighing the repulsive forces due to the positive charges of the protons (see Chapter 4). If there are 84 or more protons in the nucleus, the repulsive forces cannot be completely balanced and the nucleus is unstable. The degree of instability is indicated by the half-life of the radio-isotope. The longer the half-life, $t_{1/2}$, the greater the stability, e.g. $^{222}_{86}$Rn ($t_{1/2} = 3.82$ d) is more stable than $^{220}_{86}$Rn ($t_{1/2} = 55.6$ s), with $^{219}_{86}$Rn ($t_{1/2} = 3.96$ s) the least stable of the three naturally occurring radon isotopes. The half-lives of these three isotopes are so short that the only reason that there is any radon still present in the Earth is because it is being constantly produced by the breakdown of other radio-isotopes. All the naturally occurring heavy radio-isotopes, i.e. those with atomic numbers greater than 82 (lead), are members of a radioactive decay series (Figure 16.1). These series commence with one of the three relatively stable radio-isotopes ^{238}U, ^{235}U and ^{232}Th and terminate with ^{206}Pb, ^{207}Pb and ^{208}Pb respectively. The original concentration of ^{235}U must have been 80 times greater than it is now, with twice as much ^{238}U and a quarter more ^{232}Th. These values can be calculated from the half-lives of the isotopes. The original unstable nuclide undergoing radioactive decay is called the parent and the product is called the daughter. In a decay series, each daughter becomes the parent of the following daughter element, e.g.

^{238}U (parent) $\longrightarrow$ ^{234}Th (daughter); ^{234}Th (parent) $\longrightarrow$ ^{234}Pu (daughter)

Table 16.1 The average annual radiation dose received by the population of the UK

Source	Annual radiation dose (mSv)	(% of total)
natural sources		
radon-222	1.175	47
radon-220	0.1	4
food and drink	0.3	12
cosmic rays	0.25	10
buildings, ground, etc. (non-radon nuclides)	0.35	14
sum of natural sources	2.175	87
anthropogenic sources		
medical (X-rays, radiotherapy)	0.3	12
fall-out from bomb tests	0.01	0.40
miscellaneous (smoke detectors, TV screens, old luminous clocks, etc.)	0.01	0.40
work	0.005	0.20
nuclear-power industry	0.001	0.04
sum of anthropogenic sources	0.326	13.03
total all sources	2.501	100

etc. (Figure 16.1). The abundances of daughters/parents are kept constant by the equilibrium between the rates of formation and decay. The majority of the more than 60 naturally occurring radionuclides are members of the three decay series illustrated in Figure 16.1. These, together with ^{40}K, are mainly responsible for the natural background radioactivity on the Earth's surface.

16.1 Radon in buildings

The parent of ^{222}Rn is $^{226}_{88}Ra$ (radium) which has a half-life of 1602 years and is widely distributed in rocks, sediments and soils along with the other long-lived members of the ^{238}U series. The radioactive decay of the radium atoms, whose compounds are usually solids, produces radon gas that can diffuse through the rocks and soils into the atmosphere. If there is a building over these deposits, the radon will enter that building unless there is a gas-impermeable barrier. The quantity of radon reaching the atmosphere will depend upon the quantity of radium in the rocks and soils, how near the surface the radium is and how permeable the rocks and soils are. If the permeability is low because there are few connecting pores, fissures and cracks, the radon could have largely, or completely, decayed away before it reached the surface. With a half-life of 3.82 days, over 99% of a given emission of radon will have decayed inside a month. The immediate daughter products are very short lived alpha-particle emitters (Figure 16.1) and they are transformed into $^{210}_{82}Pb$, a beta emitter, which has a half-life of 22.3 years. The lead is adsorbed on to solid surfaces, which prevents its escape into the atmosphere. Once the radon reaches the atmosphere it and its daughters can be breathed in. The lead then becomes much more important because the solid radioactive particles are trapped in the lungs, whereas the gaseous radon is breathed out again.

(a) The uranium radioactive decay series that produces radon-222

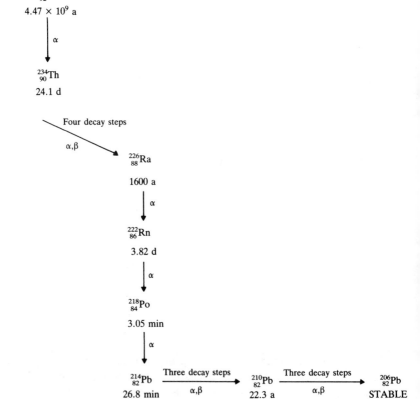

(b) The thorium radioactive decay series that produces radon-220

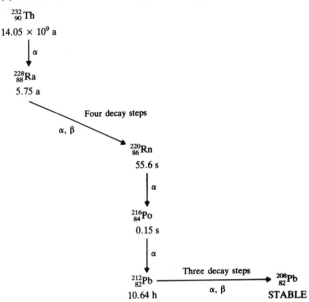

(c) The actinium (or uranium-235) radioactive decay series that produces radon-219

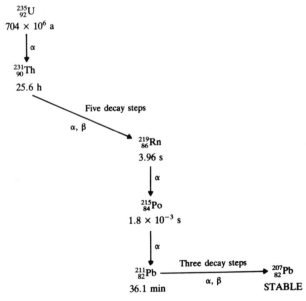

Figure 16.1 The three naturally occurring radioactive decay series that produce radon. The half-life of each nuclide is given below its symbol. Vertical arrows indicate α emission and horizontal arrows indicate β emission. Where several steps are grouped together both α and β emissions occur.

Measurements of radon concentrations in houses have been made since the 1970s. The number of houses surveyed has been increased as it became apparent that relatively high concentrations of radon were much more common than had at first been thought. The UK action level for the annual average radon concentration in houses was reduced from 400 **Becquerel** m^{-3} ($Bq\,m^{-3}$) to 200 $Bq\,m^{-3}$ in 1990 (the US action level is 150 $Bq\,m^{-3}$). It has been estimated that 70 000–100 000 homes in England will be above this level (mainly in Cornwall, Devon, Somerset, Northants and Derbyshire) and 2000 homes in Scotland (mainly in the Grampians and Highlands). The action level of 200 $Bq\,m^{-3}$ implies a lifetime risk that 1% of non-smokers breathing this air will die of lung cancer due to exposure to radon and its daughters.

The problem is probably greater now than it was in the past because houses are better insulated and have fewer draughts. However, it could be that the problem has only recently been identified because extensive radon measurements have only been carried out recently. A reduction in the number of air changes in a house will allow the radon seeping through the floor to build up, as there will be less dilution with 'fresh' air from outside the house. The daughter elements, especially ^{210}Pb, accumulate on the household dust particles suspended in the air and deposited on surfaces. Exposure to radioactivity is increased every time dust is disturbed in a room and the fine particles enter the air that is breathed. The concentration of radon outside the building drops very rapidly as it leaves the soil and winds mix the gas with air. In a house there are weaker air currents and radon levels are about 50% greater in ground-floor rooms than in first-floor rooms. The small pressure differences between the inside and the outside of buildings tend to draw in the radon preferentially from the surrounding ground. The amount of radon entering the living space of houses can be reduced by interception so that is is well below the action level. A hole filled with coarse porous material is constructed in the ground under the house. This hole acts as a sump and the gas enters this first. The radon can then be sucked out of the sump by means of a tube and a fan connected to the outside of the house. The ground floor can be further insulated from radon entry by installing a continuous sheet of heavy-duty polythene over the floor (if the house has already been built) or under the floor (if the house is being built). In older houses cracks in and around the flooring can also be sealed. In areas such as south-west England that are known to be at risk building regulations now require radon protection measures to be installed in all new buildings.

16.2 Biological effects

The properties of alpha, α, particles (helium nuclei) are such that they are only a health hazard if the emitter is in contact with living tissue. The relatively large, positively charged α particles rapidly lose their energy to surrounding atoms. This means that they do not penetrate the skin and are stopped by a few centimetres of air. However, if the α emitter is in the body, e.g. on lung tissue, all of the energy is released in a very small volume. The chances of this energy breaking chemical bonds in molecules or causing ionization or forming reactive free radicals are relatively high. All of these changes have the potential to disrupt the normal metabolism and produce changes which may either be easily reversed or affect cell replication and induce cancerous growth. For these reasons α particles are given a weighting factor that is ten times greater than that given to beta minus, β^-, particles when comparing their potential for inducing biological damage. β^- particles are electrons. They have a lower mass (approximately $1 : 7300$) and therefore travel faster and further before losing the same amount of energy as an α particle. This means that the density of radiation damage inside the body from ingested β emitters is lower than from α emitters of the same energy because the energy from the α particles is released in a smaller volume. The sievert, Sv, is a dose equivalent unit that attempts to take account of the different effects of different types of radiation on body tissues so that comparisons can be made between different radiation sources. Table 16.1 indicates the average background radiation dose that the UK population receives. The actual dose any individual will receive depends upon where they live and work. Levels of radon in the home can make a great difference, as will building materials; for example, clay bricks (containing radioactive ^{40}K) are five times more radioactive than sandstone blocks. An increase in altitude of 2000 m doubles the dose received from cosmic radiation. The medical use of X-rays has produced a significant increase in dose. The use of radio-isotopes and gamma, γ, ray irradiation, particularly for cancer therapy, can lead to very high dosages being given to individuals. A single whole-body dose of 250 mSv might cause a reduction in white blood cell count, whereas a dose of between 2 and 5 Sv is likely to kill 50% of those exposed within 60 days. More controversial are the detrimental effects caused by lower dose rates. It is generally assumed that there is a linear non-threshold dose–effect relationship in which all doses greater than zero received during a lifetime contribute to causing biological effects. There is then a finite probability of the occurrence of a detrimental effect, such as cancer, that is proportional to the dose received. It has been estimated that living in a house with a radon concentration equivalent to $20 \, \mathrm{Bq \, m^{-3}}$ gives an effective dose of $1 \, \mathrm{mSv \, a^{-1}}$. This would produce a lifetime risk of 1 in 1000 of developing cancer for non-smokers and a 1 in 100 chance for smokers. The methods by which these sorts of estimates are arrived at are of variable precision and depend upon various assumptions about the interpretation of the data upon which they are based. The larger the dose the greater the risk and the more reliable the risk estimates are likely to be.

17 Hazardous organic compounds

17.1 Synthetic organic compounds

The large-scale production of synthetic organic chemicals commenced in the middle of the nineteenth century. Initially coal was the major raw material for the production of coal-tar oil from which most of the early organic compounds were extracted or synthesized, but gradually oil and natural gas have become the favoured starting materials. With increasing knowledge of chemical structure and increasing technical ability the range of compounds synthesized has widened and become more divergent from naturally available substances. Some of the synthesized compounds such as polychlorinated biphenyls (PCBs) and chlorofluorocarbons (CFCs) were manufactured because they were very stable. It is now realized that this stability, which made them so suitable for their intended usage (e.g. PCBs as coolant–insulator fluids in electricity transformers), is causing problems when they are released into the general environment. They are very persistent because micro-organisms and other agents that normally bring about the biodegradation of natural organic materials are either unable to break down these stable molecules or can do so only extremely slowly. Similarly inorganic reaction mechanisms may be ineffective, for instance with CFCs in the troposphere (see Chapter 2), or produce unwanted products (chlorine from CFCs in the stratosphere). The concentration of these organic compounds in air, water, soil and organisms is generally very low (of the order of parts per billion or trillion). However, their persistence, possible **biomagnification** and potential toxicity is a cause of great concern.

When considering the risks posed by these synthetic organic chemicals it is useful to remember that many naturally occurring organic chemicals are also extremely toxic. Very few naturally occurring organic chemicals have been tested in the same way as the synthetic chemicals. Of the small number that have been tested using animal cancer bioassays, about half were found to be carcinogenic. The concentrations of these natural carcinogens in food are generally thousands of times greater than the concentrations of synthetic chemicals, e.g. orange juice (30 ppm); apples, potatoes (50 ppm); brewed coffee (90 ppm); lettuce (300 ppm); raspberries, strawberries (1000 ppm); comfey tea (3000 ppm). Alcohol (ethanol) is carcinogenic and its potency, as determined in rodent tests, can be compared with that of the most potent of the dioxins (TCDD, see section 17.2). If the permitted level of ethanol ingestion was set so that it would have the same deleterious effect as the acceptable daily maximum dose of TCDD, then the recommended level would be 1×10^{-4} g of alcohol per day (1 glass of wine or a half-pint of beer every 350 years!). Another naturally occurring compound, indole carbinol, reacts in a similar manner to TCDD with regard to stimulating carcinogenesis. Cabbage, cauliflower and broccoli contain 50–500 ppm of indole carbinol. One helping of broccoli gives an effective dose of this potential carcinogen that is 1500 times greater than the maximum recommended TCDD dose (these figures take into account the relative potency and lifetimes in the body of the two compounds). The widespread occurrence of relatively high concentrations of natural toxins should not be ignored when considering the extent of the risk posed by the ingestion of synthetic organic chemicals.

Tens of thousands of new compounds are synthesized each year and hundreds of new compounds are manufactured and sold in varying quantities from a few grams to thousands of tonnes. It is not possible to highlight more than a few groups that illustrate the factors that are causing concern.

17.2 Dioxins

The dioxins are two groups of compounds that are found as impurities in the manufacture of some chlorinated organic compounds or when chlorinated compounds are burnt at relatively low temperatures (less than 1000 °C). These two groups are (a) the polychlorinated dibenzo-*p*-dioxins (PCDDs) (Figure 17.1) and (b) the polychlorinated dibenzofurans (PCDFs) (Figure 17.1). There are 75 members of the PCDD group and 135 members of the PCDF group. One member of the PCDD group is also referred to as dioxin. This is 2,3,7,8-tetrachlorodibenzo-*p*-dioxin (Figure 17.1).

The toxicities of the individual members of the two groups are very different. The most toxic compounds have four chlorine atoms (tetrachloro) in the 2,3,7 and 8 positions (Figure 17.1); 2,3,7,8-tetrachlorodibenzo-*p*-dioxin (2,3,7,8-TCDD) is extremely toxic to laboratory animals (e.g. LD_{50} guinea pigs 0.6 μg kg^{-1} body mass), and is considered to be the most dangerous of all the dioxin group. The least

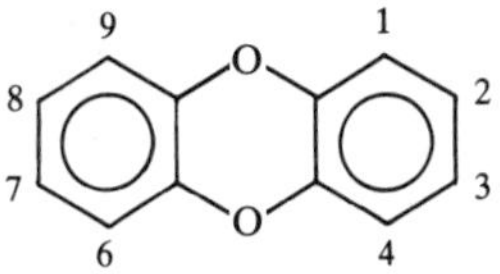

(a) Dibenzo-*p*-dioxin

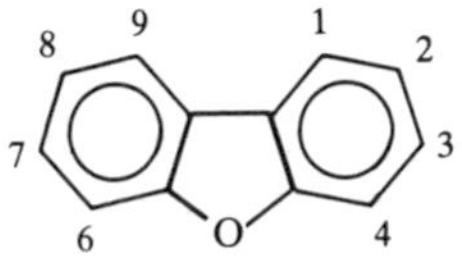

(b) Dibenzofuran

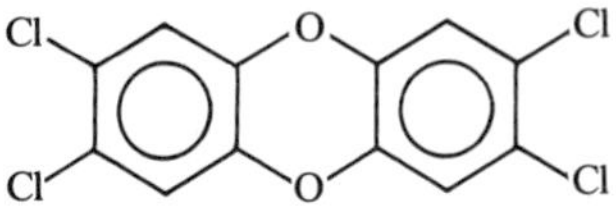

(c) 2,3,7,8-Tetrachlorodibenzo-*p*-dioxin

Figure 17.1 Structures of (a) dibenzo-*p*-dioxin and (b) dibenzofuran indicating the eight positions that can have chlorine atoms attached to form the two groups of polychlorinated compounds described as dioxins. (c) The structure of 2, 3, 7, 8-tetra-chlorodibenzo-*p*-dioxin, TCCD.

toxic compounds have one to three chlorine atoms (monochloro to trichloro) and are considered to be relatively harmless. Whilst many of the dioxins are extremely potent carcinogens for animals their effects on humans are less certain. In 1976 there was an explosion at a chemical works in Seveso, Italy, that contaminated an area of about 3 square miles (7.8 km^2) with dioxins. Many animals died but the 36 000 people in the affected area of the town survived, though 193 suffered from severe chloracne (a skin disorder). Only one case of chloracne had not cleared up 10 years later. There appears to have been no major long-term health effects. It is still possible that cancers linked to the exposure of the population to dioxins could develop because the induction period could be 20 years or more. However, there has been no convincing evidence that dioxins are a potent carcinogen for humans. The mean dose of TCDD to the children that developed chloracne was 3.1 μg kg^{-1} body weight, much higher than that which was so deadly for guinea pigs. Studies of people exposed to dioxins either from handling contaminated chemicals or due to accidental release have not yet shown dioxins to be especially dangerous. 'Background' concentrations of TCDD in human fat, where it is stored in the body, range from non-detectable to 20 pg kg^{-1}. Children with chloracne at Seveso had a mean TCDD concentration of

22 000 pg kg^{-1} (range 800–28 000 pg kg^{-1}). The half-life of dioxins in humans is about 7 years. Over 98% of human intake is via food, especially from beef and milk. Airborne dioxins are deposited on fodder and soils from where they can enter the food chain. The half-life of dioxins in soil has been estimated to be about 100 years.

The dioxin group of compounds are produced in a number of processes involving the use of chlorine and chlorine compounds. The dioxins are byproducts of these reactions. Which of the 210 dioxins are actually formed, and in what quantity, depends on the process conditions. As a consequence, the potential hazard due to these impurities also varies. Chlorophenols such as the familiar 'TCP' (trichlorophenol) are widely used as wood preservatives, bactericides, insecticides and herbicides. They are also intermediates in the production of paper, paints and textiles. PCBs are another group of chlorinated organic compounds that contain small quantities of dioxins. The combustion of chlorine-containing mixtures in municipal and industrial waste incinerators will produce dioxin compounds unless the temperature is above 1200 °C. This means that older municipal incinerators can be significant local sources of dioxins, emitting between 1000 and 7500 μg t^{-1} solid waste incinerated, depending upon its composition. Many plastics, e.g. polyvinyl chloride (PVC), contain organochlorine compounds that can be involved in dioxin production. Leaded petrol contains chlorine-containing compounds such as ethylene dichloride as a scavenger for the lead produced during combustion. Therefore, vehicle exhaust emissions and used lubricating oil also contain small quantities of dioxins. As the use of unleaded petrol increases this source of dioxins is becoming less important.

Because the various members of the dioxin group of compounds have very different toxicities, it is important to know the compound profile of the dioxin mixture as well as the total quantity of dioxins present. With 210 compounds to be considered this leads to some practical difficulties in presenting the results, as well as carrying out the analysis. One attempt to take these problems into account involves the use of the concept of the 'toxic equivalent' (TEQ) value. Each of the dioxins is assigned a 'toxic equivalent factor' (TEF). The TEF for 2,3,7,8-tetrachlorodibenzo-*p*-dioxin (2,3,7,8-TCDD) is 1 because it is considered the most toxic of the dioxins. Other compounds are assigned a figure between 0 (the mono-, di- and tri-compounds) and 1 depending upon their toxicity relative to that of 2, 3,7,8-TCDD. For example, the pentachlorodibenzo-*p*-dioxins (PeCDDs) have a TEF of 0.5 if the chlorines are in the 1,2,3,7,8 positions, but a TEF of 0 for the other PeCDDs. The pentachlorodibenzofurans (PeCDFs) also have a TEF of 0 unless chlorines are in the 2,3,4,7,8 positions

(TEF 0.5) or in the 1,2,3,7,8 positions (TEF 0.05). The quantitative composition of the mixture of dioxin compounds is determined and the result for each component is multiplied by the relevant TEF. These values are than added together to give the TEQ. The resulting single-figure TEQs allow easier comparison of the emissions from different sources. For example:

Source A:
  Emission of
        1 g 2,3,7,8-TCDD + 4 g PeCDD = 5 g dioxins/day
          TEQ = $(1 \times 1) + (4 \times 0) = 1$ g/day
Source B:
  Emission of
        0.75 g 2,3,7,8-TCDD + 2 g 2,3,4,7,8-PeCDF = 2.75 g dioxins/day
          TEQ = $(0.75 \times 1) + (2 \times 0.5) = 1.75$ g/day

This indicates that though source B is emitting a lower quantity of dioxins, the danger from this source is potentially greater. However, it must be remembered that (a) the TEF values are only approximations for the relative toxicities of the individual compounds, and (b) the simple addition of the values may not be a good representation of the combined effects of a complex mixture of dioxins.

Dioxins are contaminants in chlorophenols, halogenated herbicides such as 2,4,5-trichlorophenoxyacetic acid (2,4,5-T) and 2,4-dichlorophenoxyacetic acid (2,4-D), and in polychlorinated biphenyls (PCBs). They are, therefore, found wherever these chemicals have been used or disposed of. They are further distributed by incomplete combustion of chlorinated organic compounds. Very sensitive methods of analysis have been developed so that it is possible to identify dioxins at extremely low concentrations. We know that dioxins are present at these very low levels (e.g. 9 pg TCCD kg^{-1} milk in the UK), but do not know what the safety level is. It has been estimated that the present human exposure level TEQ is about 1 pg kg^{-1} body weight per day with about 30% of this as TCDD.

17.3 DDT and related compounds

The insecticidal effects of 1,1,1-trichloro-2,2-di-(4-chlorophenyl)-ethane, originally called dichlorodiphenyltrichloroethane (i.e. DDT, Figure 17.2) were first exploited in the 1940s. It was used in high concentrations throughout the world, especially where malaria was endemic. It has a relatively low toxicity for humans and it was extremely effective. In the early 1960s concern was expressed about the persistence of DDT and some of its degradation products, DDE

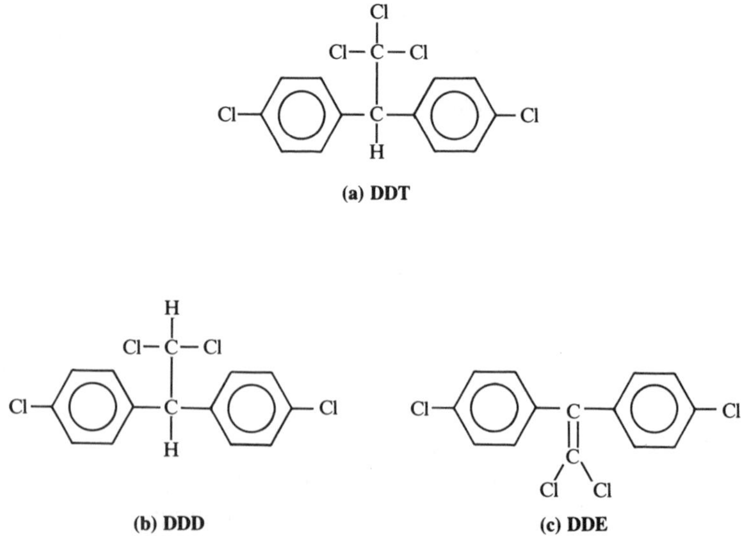

(a) DDT

(b) DDD

(c) DDE

Figure 17.2   Structures of (a) 1,1,1-trichloro-2,2-di-(*p*-chlorophenyl)ethane, DDT; (b) 1,1-dichloro-2,2-di-(*p*-chlorophenyl)ethane, DDD; (c) 1,1-dichloro-2,2-di-(*p*-chloro-phenyl)ethene, DDE.

and DDD (Figure 17.2), together with evidence that these compounds were building up in some organisms. For example, in an area sprayed with DDT to control Dutch Elm disease, the concentration of DDT in robins was 20 to 30 times higher than in the earthworms that they ate. There was a 70% decline in robin population in the sprayed area. Egg-shell thinning and the failure of fledglings to develop, especially amongst birds of prey, have been related to the presence of chlorin-ated pesticides such as DDT in their diet.

The use of DDT and related products was banned in the USA in 1972 and other developed countries later, but it is still widely used (over 40 000 tonnes per year) in developing countries where its low cost and low human toxicity are thought to outweigh its disadvan-tages. The first cases of insects becoming resistant to DDT were noted as early as 1948. As time has passed more and more of the target groups have developed resistance. The persistence of DDT and its wide-scale use encouraged the development of resistant strains of insects and other pests. The DDT kills off the susceptible individuals in the pest population and also the natural predators. Any naturally resistant pests are then able to develop rapidly because of the lack of natural competitors. The persistence of compounds such as DDT keeps down the susceptible population and allows the tolerant indi-viduals to become dominant. The resistant strains generally have some biochemical mechanism that enables them to detoxify the insecticide. For example, DDT can be detoxified by enzyme conversion to DDE (Figure 17.2), a compound that has a very low toxicity for insects. DDT affects the nervous system by interfering with the sodium balance in nerve membranes. The skin of animals prevents intake through absorption, but insect cuticles have a high permeability. This is why DDT appeared to be so safe to use. However, the persistence of DDT in soils and on foliage meant that it could enter food chains and be constantly recycled. The preferential solubility of DDT in fats (it is described as being lipophilic) meant that it was stored in fatty tissue and not readily excreted. This leads to **biomagnification**. The toxic effects on the organism may result after a relatively long time period either through levels rising above a threshold that induces adverse effects or because some stress event causes the chemical to be released from the fat.

17.4 Polychlorinated biphenyls, PCBs

The polychlorinated biphenyls (PCBs) are a group of compounds that contain from 1 to 10 chlorine atoms attached to various carbon atoms in the biphenyl molecule (Figure 17.3). There are 209 possible members of the PCB group and industrial processes produce various mixtures whose compositions depend upon the conditions of manufacture. In addition to the various members of the PCB group the commercial products usually contain small quantities of dioxins as impurities.

The PCBs are chemically stable and have a high electrical resistance. These properties made them appear to be ideal components for use in electricity transformers (which lower or raise the voltage of power lines) and capacitors (which help maintain constant voltage during power transmission). They have also been widely used as

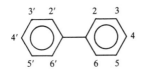

Figure 17.3 Structure of the biphenyl molecule. Each of the numbered positions can have a chlorine atom attached to it. The resulting polychlorinated biphenyls, PCBs, can have from one to ten chlorine atoms per molecule depending upon the conditions under which they are produced.

additives in lubricating and cutting oils, hydraulic fluids, heat-transfer fluids, etc.

Production of PCBs came to an end in most countries in the late 1970s. In the mid-1960s analytical methods that were capable of detecting small concentrations of PCBs became available. It was soon clear that PCBs were being transferred from the industrial and urban areas of use to places like the Arctic and Antarctic that were far removed from any local source. Of even greater concern were the findings that the PCBs were accumulating in animal fatty tissues. PCBs were found to be transferred particularly efficiently through aquatic food chains so that the highest concentrations were found in marine mammals such as seals and whales ($5-50$ mg kg^{-1} blubber), otters ($5-200$ mg kg^{-1} fat) and fish-eating birds. The concentration in the body fat may be 10 million times greater than the concentration in the water in which the mammals live. There appears to be a strong link between high concentrations of PCBs in body fat and low rates of reproduction in animals such as seals, beluga whales and otters. However, high levels of PCBs are usually also associated with high levels of other organic chemicals, so it is not clear whether it is one chemical (which may or may not be a PCB) or a mixture of chemicals that are responsible for the effects. Also, because the PCBs are mixtures of different compounds, varying effects may be due to differing mixtures. There is a weaker link between high PCB concentrations and damage to the immune system of marine mammals. Relatively high levels of PCBs ($10 \mu g$ dm^{-3}) may be found in human-breast milk. There have been suggestions that the higher PCB concentrations may cause reductions in reproductive efficiency and impair the learning ability of children, as well as having potential carcinogenic activity. The evidence is not clear because of the low concentrations and the long time period over which such effects may occur.

Disposal of PCBs can be achieved by high-temperature incineration (temperature greater than 1200 °C for more than 2 seconds in the presence of excess oxygen). This process is feasible for industrial and commercially held stocks of PCB-containing equipment and contaminated liquids. It is not feasible for contaminated solids such as soils, if large tonnages are involved. Even high-temperature incineration can cause problems unless the equipment is carefully operated and maintained. There are possibilities of producing dioxins and not completely destroying all of the PCBs unless the correct combustion conditions are fully achieved. Costs are very high – several thousand pounds per tonne! Large quantities of PCBs have been placed in landfill sites from which they can escape via the atmosphere or through ground water. Much work has been carried out on the development of micro-organisms that will breakdown PCBs *in situ*. Unfortunately, the great variability in natural conditions and competition from other micro-organisms has prevented the transfer of successful laboratory-tested biotechnology systems to natural systems in a reliable manner.

A total of about 1.2 million tonnes of PCBs were manufactured throughout the world before production ceased. Over 30% of this output has been estimated to have been released to the environment, mainly to landfill sites. Probably half of the PCBs is awaiting disposal after use. Provided all of these stocks are successfully destroyed, probably by incineration, the global levels of PCBs should remain relatively constant over the short term. There should then be a slow decline as the compounds are slowly decomposed and/or buried in sediments. The wide distribution of equipment containing PCBs, the poor storage of PCB-contaminated oils and the high cost of effective incineration make it unlikely that there will not be further releases of PCBs to the global environment.

17.5 Polynuclear aromatic hydrocarbons, PAHs

The polynuclear (or polycyclic) aromatic hydrocarbons contain two or more fused aromatic rings (Figure 17.4). Eight of the PAHs are considered to be possible or probable carcinogens. The compounds are a byproduct of incomplete combustion. Natural sources include forest fires, prairie fires and volcanic eruptions. Anthropogenic sources include motor vehicles (particularly diesel engines), coking ovens, asphalt manufacture, fossil-fuelled furnaces, cigarettes and barbecues. PAHs may be formed from any hydrocarbon source if the conditions of incomplete combustion due to lack of oxygen are met.

Chrysene

Benzo(a)pyrene

Figure 17.4   Structures of chrysene and benzo(a)pyrene, two examples of the polynuclear aromatic hydrocarbon, PAH, group of compounds.

Straight-chain saturated hydrocarbons, such as methane or *n*-octane, are least likely to form PAHs whereas cyclic hydrocarbons, as found in diesel and fuel oil, are more likely to produce PAHs. The closer the structure of the hydrocarbon to that of the PAHs the easier it is for PAHs to be formed.

The increasing use of fossil fuels has increased the exposure of humans to PAHs. In the UK, before the introduction of the Clean Air Act (1956), domestic coal fires were a major source of PAHs which tended to be adsorbed on to the surface of soot and smoke particles. The relationship between soot and cancer in chimney sweeps was recognized in the nineteenth century. Urban levels of PAHs, having fallen following the introduction of the Clean Air Act, have risen again, especially because of the emissions from poorly maintained diesel-engined vehicles.

PAHs such as benzo(a)pyrene appear to exhibit low acute toxicity but have very significant chronic toxicity, i.e. a single large dose will not cause immediate adverse effects but continuous low doses will probably induce cancer. For non-smokers, food is normally the main source (over 95%) of the carcinogenic PAHs (average daily intake about $3 \mu g$) unless they spend a lot of time in the presence of smokers or in polluted urban atmospheres. In these cases their intake may be increased by another $2-5 \mu g\,d^{-1}$. Smoking 20 unfiltered cigarettes a day will also increase the daily intake by $2-5 \mu g$. Concentrations of PAHs in food vary quite considerably with unwashed leafy vegetables $(20-40 \mu g\,kg^{-1})$, unrefined grains $(10 \mu g\,kg^{-1})$, charcoal-broiled and smoked meats and fishes $(10-20 \mu g\,kg^{-1})$ being major potential sources. Leafy vegetables and grains are contaminated by surface adsorption of atmospherically deposited PAHs. The route of ingestion needs to be taken into account when looking at potential risk. Generally intake via the lungs, as with cigarette smoke and atmospheric contamination, leads to a greater percentage absorption into the body than does contamination of solid food. The carcinogenic properties of PAHs are believed to be due to metabolites formed in the body as the compounds are broken down.

There are great difficulties in deciding the degree of risk posed by the very low concentrations of PAHs or other chemicals. It is known that chemicals show an effect at a particular concentration that is generally much higher than the concentration found in normal environmental samples. Are the same effects shown at the lower levels? How can the cause of cancer, say, be identified when the population is exposed to a large number of possible, and different, trigger chemicals that may be of natural or anthropogenic origin? In addition there may be synergistic effects between these chemicals which can considerably increase the risks posed by the individual chemicals separately. The long gestation time between exposure and clinical appearance of the cancer may well make it impossible to be certain of the cause. The outstanding work of analytical chemists in developing methods for the recognition and determination of extremely low concentrations of various elements and compounds has made us aware of their presence. Unfortunately our knowledge of toxicology has not kept pace and we are not able to quantify the risk that these low levels of chemicals pose.

Appendix
Chemical compound formation

Atoms consists of a small nucleus (radius about 10^{-14} m) containing protons, each of which has a positive charge and a mass of about 1 atomic mass unit (1.67×10^{-24} g), and neutrons, each of which has no charge and a mass similar to that of a proton. An **element** is characterized by the number of protons in its nucleus. This number is called the **atomic number**. Each oxygen atom (atomic number 8) contains eight protons; each nitrogen atom (atomic number 7) contains seven protons. If an oxygen nucleus lost a proton it would then be a nitrogen nucleus.

A change in the number of protons leads to a change in the chemical properties (and name) of an atom. The number of *neutrons* in a nucleus does not so directly affect the chemical properties of an atom; a given element may have atoms with different numbers of neutrons present in their nuclei. These atoms, with the same numbers of protons, the same chemical properties but different numbers of neutrons, are called **isotopes** of the element.

Outside the nucleus there is a much larger volume of space (approximate radius 10^{-10} m) that contains electrons. Electrons have a negative charge of the same size as the positive charge on a proton, but are much smaller in mass (0.91×10^{-27} g) than either the proton or the neutron. A free atom in its lowest energy state (its most stable state) is electrically neutral: the number of electrons outside the nucleus is the same as the number of protons inside the nucleus. If the atom loses an electron, it will be positively charged (electron deficient); it will then be called a **cation**. If an atom gains an electron it will be negatively charged (excess electrons) and will then be called an **anion**.

It has been discovered that each electron in an atom has a distinctive energy, different from that of each of the others. However, these energies can only have certain specified values (Figure A.1). The chemical properties of atoms (and also molecules and ions) are determined by the energies of the electrons they contain. As is indicated in Figure A.1, the energy levels form groups whose members differ in energy by small amounts, e.g. 2s and 2p form a group. These groups

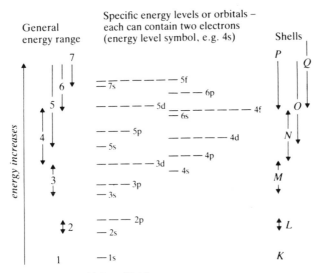

The lower energy orbitals are filled first.
The lower the energy of the electrons the more stable the structure.

Figure A.1 The arrangement of electron energy levels in atoms.

of levels with similar energies are frequently called **shells**. The various shells are often referred to by the letters *K, L, M, N*, etc., that were introduced before many of the detailed features of atomic structures were known. It is the distribution of the electrons between the levels in the outermost (highest-energy) shell that largely determines the chemical properties of the atom.

The **periodic table** (Figure A.2) is a method of arranging the elements in order of increasing atomic number so that those elements with similar electronic structures, and hence similar chemical properties, are grouped together. Elements in the same vertical groups have the same outer (**valence**) shell structure, e.g. ns^1 for lithium, sodium, potassium, rubidium, caesium and francium; and $ns^2 np^4$ for oxygen, sulphur, selenium, tellurium and polonium. (Note that the superscript indicates the number of electrons in a particular energy level.) Elements in the same period have valency electrons of the same general energy, though the number of electrons increases from left to right across the table. The group helium, neon, argon, krypton, xenon and radon is called the 'noble' or 'inert' gases; its members have a particularly stable electronic structure of full energy levels, and rarely react.

electronegativity increases (→)

electronegativity increases (↑)

Period \ Group	I	II											III	IV	V	VI	VII	O
1									1 H									2 He
2	3 Li	4 Be											5 B	6 C	7 N	8 O	9 F	10 Ne
3	11 Na	12 Mg											13 Al	14 Si	15 P	16 S	17 Cl	18 Ar
4	19 K	20 Ca	21 Sc	22 Ti	23 V	24 Cr	25 Mn	26 Fe	27 Co	28 Ni	29 Cu	30 Zn	31 Ga	32 Ge	33 As	34 Se	35 Br	36 Kr
5	37 Rb	38 Sr	39 Y	40 Zr	41 Nb	42 Mo	43 Tc	44 Ru	45 Rh	46 Pd	47 Ag	48 Cd	49 In	50 Sn	51 Sb	52 Te	53 I	54 Xe
6	55 Cs	56 Ba	57* La	72 Hf	73 Ta	74 W	75 Re	76 Os	77 Ir	78 Pt	79 Au	80 Hg	81 Tl	82 Pb	83 Bi	84 Po	85 At	86 Rn
7	87 Fr	88 Ra	89** Ac															

* lanthanide series	58 Ce	59 Pr	60 Nd	61 Pm	62 Sm	63 Eu	64 Gd	65 Tb	66 Dy	67 Ho	68 Er	69 Tm	70 Yb	71 Lu
** actinide series	90 Th	91 Pa	92 U	93 Np	94 Pu	95 Am	96 Cm	97 Bk	98 Cf	99 Es	100 Fm	101 Md	102 No	103 Lw

Figure A.2 The periodic table of the elements showing the general trends in electronegativity changes.

Though individual atoms will normally have their electrons in the lowest possible states available to them, combining together two or more atoms may allow the electrons to rearrange themselves so that there is an overall reduction in energy of the electrons (and of the atoms) in the new combined state. For instance, the combined energy of two separate atoms of oxygen is greater than the combined energy of the two atoms linked together in the dioxygen molecule. The formation of dioxygen (Eqn A.1) is thus favoured. However, the combined energy of two dioxygen

$$O + O \longrightarrow O_2 \qquad\qquad (A.1)$$

molecules is lower than the energy of four oxygen atoms combined together in a tetraoxygen (O_4) molecule. Therefore neither reaction A.2 nor A.3 occurs.

$$4O \longrightarrow O_4 \qquad\qquad (A.2)$$

$$O_2 + O_2 \longrightarrow O_4 \qquad\qquad (A.3)$$

Chemical reactions involve changes in the energy of the electrons; they do not involve changes in the energy of the nucleus.

Examination of the compounds formed by oxygen with carbon, nitrogen and sulphur has shown that they, like dioxygen, consist of molecules with the atoms held together by the sharing of electrons, i.e. they have **covalent bonds**. However, close study shows that atoms of different elements have different degrees of attraction for the shared electrons. Maybe we should not be surprised by this, when we consider the differently sized positive nuclear charges of the different elements, and the different electron energy distribution associated with each. Of all the elements, fluorine has the strongest attraction for the shared electrons; oxygen has the next greatest attraction. **Electronegativity** is defined as the power of an atom in a molecule to attract electrons to itself. The order of electronegativities for a number of elements is:

$$F > O > Cl > N > S > C \simeq H > Si > Al > Mg > Ca > Na > K$$

The elements that are more electronegative than hydrogen form compounds with oxygen that are gases at 293 K (normal room temperature), whereas the elements that are less electronegative than hydrogen form solid compounds with oxygen.

Consider what happens to the shared electrons in a molecule as the relative electronegativities of the atoms change (Figure A.3). When the electronegativities are the same, A–A (e.g. O with O), the electrons are evenly shared in covalent bonds and mainly occupy the

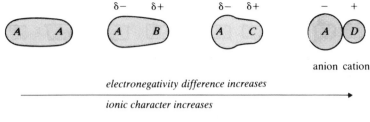

order of electronegativity: $A > B > C > D$

Figure A.3   The effect of changes in electronegativity on the electron distribution between two atoms bound together.

space between the two nuclei. The bonding electrons are said to be 'localized'. One result of this is that molecules containing three or more atoms will have a specific shape depending upon the relative localized positions of the bonding electrons (Figure 3.5). The final geometry of the molecule, which is generally the arrangement with the lowest energy, is determined by various interactions between bonding and non-bonding electrons and by inter-nuclear repulsions.

As the electronegativity difference increases, the charge imbalance increases, and an **electrostatic attraction** develops that is non-localized. The greater the size of the electrostatic force, the higher the melting and boiling points of the compound and the greater its ionic character. With large differences in electronegativity, there can be complete transfer of an electron to form a negatively charged anion and a positively charged cation. These changes in numbers of electrons mean that anions are usually larger than cations. The system reaches its lowest energy state, and greatest stability, when each cation is surrounded by the maximum number of anions and each anion is surrounded by the maximum number of cations (Figure A.4). This maximum depends upon the charges and sizes of the ions involved. The result is a structure that can be compared to sets of spheres packed together as closely as possible, with oppositely charged ions being in contact and ions of like charge being out of contact. The relative sizes of the anions and cations determine the packing arrangement. The relative sizes can be compared by means of the radii of the ions, and the geometrical arrangement can be predicted from the radius ratio of the ions, where

$$\text{radius ratio} = \frac{\text{radius of cation}}{\text{radius of anion}} \qquad \text{(Table A.1)}$$

We now have a picture of compounds formed by elements of very different electronegativities that do not contain simple discrete molecules, but that are extended three-dimensional arrays of charged

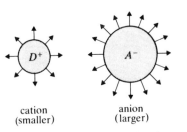

cation
(smaller)

anion
(larger)

(a) Attractive forces act in all directions

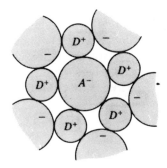

(b) Anion surrounded by cations

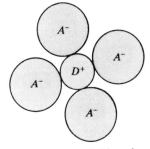

(c) Cation surrounded by anions

Figure A.4 Spatial relationships between ions in compounds.

spheres. These are described as ionic compounds, and the non-directional, electrostatic attractive forces are called **ionic bonds** (or **electrovalent bonds**).

The differences in electronegativities between the elements can be related to their positions in the periodic table (Figure A.2). If we ignore the noble-gas group we find that electronegativity increases from left to right across the periods and from bottom to top up the groups, so that the elements at the top right-hand side are the most electronegative and the elements at the bottom left-hand side are

Table A.1 Relationships between the radius ratio (radius of cation:radius of anion), the co-ordination number of the cation (number of surrounding anions) and the geometrical arrangement of the structure

Radius ratio	Co-ordination number	Shape
0.155	3	planar trigonal
0.225	4	tetrahedral
0.414	6	octahedral
0.732	8	simple cubic
1.0	12	close packed

the least electronegative, i.e. most electropositive. Therefore by comparing the positions of two elements in the periodic table, we can gain a good idea of how ionic or covalent a bond between them will be. A large difference, say more than five places, will indicate ionic character, whereas a smaller difference will indicate covalent character. There are, of course, some compounds that show intermediate character.

Glossary

accuracy The difference between the observed value and the true value. A statement of accuracy can only be made if the true value is known: in environmental measurements it is often very difficult to know what the true value is.

acid rain (acid precipitation) Rain (precipitation) whose pH is less than 5.6, the normal equilibrium value for carbon dioxide and water. The pH is reduced due to the presence of acids (mainly sulphuric and nitric) produced by the combustion of fossil fuels or released by natural events such as volcanic eruptions.

aerobic conditions Environmental conditions in which dioxygen is present.

anaerobic conditions Environmental conditions in which dioxygen is absent.

anion A negatively charged chemical species. In the simplest case an atom has gained one or more extra electrons, e.g. chloride, Cl^-, oxide, O^{2-}; but the anion may contain *several* atoms such that the group has an overall negative charge, e.g. nitrate, NO_3^-, acetate, CH_3COO^-. If the anion is in a mobile state (in solution or molten) it will be attracted to the anode (positive electrode) when an electrical potential difference is applied. No movement occurs when in a solid.

aromatic compound (aromatization) A planar cyclic organic compound in which there is a delocalized bonding-electron system. The system is composed of carbon atoms which apparently have alternating single and double bonds, $C-C=C$; however, the chemical properties show that the molecules have greater stability than this alternating structure should give. One pair of electrons in each double bond is thought to be spread over the whole aromatic portion of the molecule, $C\cdot C\cdot C$.

autotrophs Organisms that can produce organic molecules from inorganic sources. The majority are **photosynthetic** autotrophs and utilize the energy in sunlight, but some can obtain all their required energy from inorganic chemicals and are called chemosynthetic autotrophs.

Avogadro's number 6.023×10^{23}. The number of atoms in 12.00 g of $^{12}_6C$; or, more generally, the number of particles in one **mole** of any substance.

Becquerel The SI unit of activity of nuclides that are unstable and, therefore, radioactive.

$$1 \text{ Becquerel } (1 Bq) = 1 \text{ disintegration per second}$$

This means that 1 Bq represents one nucleus disintegrating each second. It is a much smaller unit of activity than the previously used unit which was the

Curie (Ci), where

$$1 \text{ Curie} = 3.7 \times 10^{10} \text{ Bq} = 3.7 \times 10^{10} \text{ disintegrations per second}$$

biomagnification   The increase in concentration of a chemical that occurs in a living organism because the rate of excretion is lower than the rate of ingestion. This can result in the concentration of a compound being very much higher in the organism than in the environment in which the organism lives. A further effect may be that there is an increase in concentration up the **food chain** as each succeeding member feeds on organisms containing increasingly high quantities of the chemical.

biota   A general term denoting all living matter. **Biomass** is mass of biologically derived material, and this may be subdivided into living biomass and dead biomass.

buffer   A solution that shows only a small change in pH when an acid or alkali is added to it (see Chapter 4).

catalyst   A substance that accelerates the rate of a chemical reaction, but which itself remains unchanged at the conclusion of the reaction. The catalyst reduces the energy barrier that holds back a reaction. It does not increase the ultimate yield, but it does shorten the time taken to obtain the products. The catalyst can be continuously reused unless it is 'poisoned'.

cation   A positively charged species generally formed by the loss of one or more electrons from an atom of a metal, e.g. sodium, Na^+, calcium, Ca^{2+}. If in a mobile state, a cation is attracted to the cathode (negative electrode) when an electrical potential difference is applied.

cation-exchange capacity   A measure of the number of negatively charged sites that are capable of reversibly binding cations. In soils these sites are associated mainly with organic matter and clays (see Chapter 12).

cell   The fundamental unit of structure of living organisms. The major division into two cell types is based on whether (a) a separate nucleus containing the genetic material is present (the *eukaryotes* – includes all modern multi-celled organisms), or (b) the genetic material is not separated by a membrane from the rest of the cell contents (the *prokaryotes*).

complex ion or **molecule**   A central atom or ion (generally a metal) attached to more atoms or groups of atoms than would be expected from the charge on the central atom or ion, e.g. Fe^{3+} in $[FeCl_4]^-$, tetrachloroiron(III)ate. The atoms etc. surrounding the central atom or ion are called *ligands*. Some ligands contain more than one atom that can be attached to the central atom or ion and these tend to form stronger complexes called *chelates*. For example, ammonia (NH_3) has one binding atom, the nitrogen, and can form the complex $[Cu(NH_3)_6]^{2+}$ with copper. Glycine ($NH_2CH_2CO_2H$), on the other hand, can bind through the nitrogen *and* through one of the oxygens: it can therefore form a chelation complex with copper, $[Cu(NH_2CH_2COO)_3]^-$. The complex should always be surrounded by square brackets [ ] when the formula is written.

consumers   see **food chain** and **heterotrophs**.

covalent bond   The chemical bond formed between two or more atoms by the sharing of electrons. Each covalent bond can be considered to contain two electrons. The bonding is directional in space, and a molecule or group containing at least three atoms bound together by covalent bonds will have a characteristic shape. There may be up to three covalent bonds joining two atoms.

crystal lattice   The geometrical structure in which the positions of the atoms, ions or molecules in a crystal are represented by points. The crystal lattice is made up of repeating units of lattice points, each of which is called a unit cell. The arrangement of atoms, ions or molecules in the crystal is influenced by the relative sizes of, charges of, and degrees of covalent bonding between the components.

depression of freezing point   The lowering of the freezing (melting) point of a solvent that occurs when a substance is dissolved in the solvent to form a solution. The larger the concentration of substance dissolved in the solvent, the greater is the depression of freezing point. This phenomenon is made use of when adding antifreeze to radiator water and when salting roads to prevent ice formation.

Dobson units   A measure of the quantity of ozone in the atmosphere. One Dobson unit (1 DU) is the amount of ozone in the vertical column of air that, if concentrated at sea level at normal temperature and pressure, would occupy 0.1 mm. The normal amount of ozone in the atmosphere gives values in the range 300 to 350 DU.

Donnan membrane   A membrane that allows some solutes (dissolved species) as well as solvent to pass through. Other solutes are prevented from passing through. A true *semipermeable membrane* will only allow solvent to pass through. Most natural membranes, such as cell walls, are Donnan membranes rather than semipermeable membranes (see Figure 10.4).

dry deposition   The removal of material from the atmosphere by contact with solid surfaces on which the material falls or against which it blows. The surfaces are much more effective as removal agents if they are wet.

electronegativity   The power of an atom in a molecule to attract electrons to itself. Non-metals have higher electronegativities than metals. When writing a formula it is usual to place the *less* electronegative elements first followed by the more electronegative elements, e.g. potassium sulphate, K_2SO_4.

electronic structure   The arrangement of electrons in the various available energy levels (also termed *orbitals*) in atoms or molecules. The electrons tend to occupy the lowest available energy levels and it is the rearrangement of the electrons amongst the various energy levels that is responsible for the energy changes associated with chemical reactions.

electrostatic attraction   The attraction that exists between a positive and negative electrical charge.

endothermic   Absorbing heat from the surroundings by a system undergoing a chemical reaction or some other change. A positive **enthalpy** change is described as being endothermic.

energy   The capacity to do work. An object or system can possess potential and kinetic energy. *Potential* energy arises from a system's composition (e.g. chemical energy, thermal energy) and also from the system's relative position (e.g. water at the top of a waterfall compared with water at the bottom). *Kinetic* energy arises from the motion of an object. The total amount of energy in a system and its surroundings remains constant.

enthalpy, *H*   A measure of the total energy in a system at constant pressure. The heat absorbed or evolved in a chemical reaction at constant pressure is the enthalpy change. It is possible to measure enthalpy changes, ΔH, but not the absolute enthalpy (see **entropy** and **free energy**).

entropy, *S*   A measure of the 'unavailable' energy in a system. In an isolated system any spontaneous process leads to an increase in entropy, i.e. the new arrangement is more probable than the old and there is a lower degree of order. The energy used to bring about this increase in entropy is not available for doing external work. Reduction in entropy can only occur in a non-isolated system when energy is imported, but the part of the system that exported the energy has to undergo a change in entropy to compensate for the export (see **enthalpy** and **free energy**).

enzyme   A protein-based molecule that has the ability to increase the rate of some chemical reactions by acting as a **catalyst**. As with catalysts in general, the enzyme only increases the amount of product produced in a given short period of time; it does not increase the final theoretical yield (see section 4.3).

equilibrium constant   The constant that expresses the relationship between the concentration of products and reactants in a chemical reaction that has reached equilibrium. The higher the value of the equilibrium constant for a given reaction, the higher the proportion of products there will be when equilibrium has been reached (see **equilibrium reaction**).

equilibrium reaction   A chemical reaction that will proceed until the concentrations of reactants and products are in equilibrium. The quantity of reactants being converted to products is equal to the quantity of products being reconverted to reactants. Provided that the system remains closed (nothing is added or removed) the reaction will proceed without perceptible change. The only way to use up more reactants is to change the conditions. This may involve the removal of one of the products from the reaction system.

errors   Differences between true values and observed or measured values. Errors may be divided into random errors and systematic errors. *Random errors* are fluctuations that only fit a pattern to the extent that they can be described by a 'Gaussian' or 'normal' distribution curve (Figure 1.7a). If enough values are measured there is no bias in the results and the **mean** (the average of all results) is identical to the *mode* (the most common value) with both being equal to the true value. The size of the random errors can be evaluated by taking a sufficiently large number of measurements (see **standard deviation**). *Systematic errors* are those that bias the results in one direction due to factors such as (a) incorrect use of, or faults in, a measuring instrument, or (b) incorrect assumptions in any theory used to calculate a quantity from the experimental results (e.g. assuming that it is the concentration of an element that describes its distribution in rocks when it is actually the *logarithm* of the concentration: Figure 1.7). Systematic errors can only be reduced by employing the best possible techniques and instruments. They will only show up if the true value is known, and are not revealed by statistical tests of precision.

exothermic Evolving heat to the surroundings by a system undergoing a chemical reaction or some other change. A negative enthalpy change is described as being exothermic.

fault A fracture surface in the Earth's crust along which there has been a relative displacement of the rocks on either side. Movement of the rocks along the fault occurs when the forces being applied overcome the resistance to displacement and an earthquake results.

fermentation An enzyme-controlled reaction causing the breakage of bonds in organic molecules and the release of energy; it does not involve dioxygen in the electron-transfer process.

ferromagnesian silicates Silicate minerals containing iron and magnesium in varying proportions. The minerals are dark coloured.

food chain A scheme of feeding relationships that unites the member species of a biological community. In the case of a grazing food chain, the first member (or first *trophic* level) is some **autotrophic** organism such as a plant; this is then consumed by a herbivore (second trophic level), and then successive sets of predators (carnivores) complete the chain. The herbivores and carnivores are heterotrophic organisms. *Saprophytic* (or *detritus*) food chains derive their energy and chemicals from dead matter and the interrelationships of the members of this food chain are not those of prey–predator.

free energy The Gibbs free energy, G, is a measure of the available energy in a system at constant pressure. As with **enthalpy** is it only *changes* in free energy, ΔG, that can be measured. The relationship between enthalpy, H, **entropy**, S, and free energy, G, is given by the following relationship:

$$G = H - S T$$

'available energy' = 'total energy' − 'unavailable energy'

where T is the absolute temperature (degrees Kelvin). If a reaction shows a decrease in free energy (i.e. if the value of ΔG is negative) the reaction will proceed spontaneously because the products are more stable (in a lower energy state) than the reactants; energy will be released. However, the term 'spontaneous' just means that the reaction is thermodynamically favoured: there may in practice be a kinetic barrier that prevents the reaction occurring at a measurable rate.

free radical A species containing a single unpaired electron. Such species tend to be chemically reactive.

heterotrophs Organisms that cannot utilize only inorganic molecules to synthesize organic molecules. They obtain their energy from organic molecules. Those heterotrophs that obtain their energy from living organisms are called *consumers*, whereas those that obtain their energy from dead organisms or dispersed organic compounds are called *decomposers*.

inorganic   See **organic**.

ion pair   Two single oppositely charged ions that are associated together as a distinct unit in solution. The charge of the ion pair is the sum of the charges on the two individual ions and therefore may be negative, zero or positive.

isomorphous   The replacement of one atom or ion in a crystal structure by another atom or ion of similar size in such a way that the geometrical arrangement of the crystal lattice remains intact. The charges on the exchanged ions may or may not be the same.

isotopes   Atoms that have the same atomic number (number of protons) but different numbers of neutrons in the nucleus. The chemical properties of isotopes are the same, but there are slight differences in physical properties and rates of reaction.

kinetic energy   The energy associated with the motion of a body:

$$\text{kinetic energy} = 0.5 \times \text{mass of body} \times \text{velocity}^2$$

kinetics   Information about the rates of reactions. The rate of reaction may or may not be dependent upon the concentrations of all the reactants. Differences arise because of differences in reaction mechanisms (how the reactants come into contact and how bonds are broken or formed). Whereas thermodynamics provides information about energy changes associated with chemical reactions and indicates the probability that a reaction may occur, it does not tell whether the reaction will occur at a measurable rate. For example, diamonds are thermodynamically unstable compared to graphite at 20 °C, but they are kinetically stable because of the great amount of energy required to rearrange the carbon atoms from the diamond lattice to the graphite lattice. In general, the higher the temperature the more rapidly a reaction will occur.

mean   The average value of a set of results obtained by dividing the sum of the values by the number of results.

Mohorovičić discontinuity   The region below the Earth's crust in which there is a sharp increase in the velocity of shock waves. This increase is thought to be associated with changes in composition and mineral phases. The zone is taken to be the base of the crust and lies at about 6 km below the ocean bottom and 40 km below the continental surface.

molar concentration   The concentration expressed in terms of the number of moles of a substance present. The *mole* is the amount of substance containing **Avogadro's number** of particles of that substance. In general, the value of the mole is obtained by adding together the atomic masses of all atoms in the chemical formula and expressing the result in grams, e.g. 1 mole of nitrogen oxide, NO, contains $(14 + 16)\,g = 30\,g$ of NO.

molarity, M The number of moles of substance contained in 1 litre (1 dm^{-3}) of solution.

molecule A combination of atoms in fixed numbers to give a discrete structural entity with no overall electrical charge. The molecule may contain atoms of only one element, e.g. dioxygen, O_2, or atoms of several elements, e.g. trichloromethane, $CHCl_3$.

monomers The individual units that are linked together to form polymers, e.g. amino acids in proteins.

non-polar Having no electrical charge. Used to describe many covalent organic molecules. Non-polar compounds are insoluble in polar solvents such as water but dissolve in non-polar liquids such as benzene and trichloromethane.

normal distribution A distribution of frequency of occurrence of measured values that follows a Gaussian distribution curve (see Figure 1.7a).

organic/inorganic A distinction that was first introduced to distinguish between those compounds and molecules that were thought to be formed only by living organisms (i.e. organic ones) and those in which living organisms played no part (i.e. inorganic ones). Nowadays organic compounds are those containing carbon, usually hydrogen, and varying amounts of other elements. It is not a distinct class as the boundary between the two groups is not now clearly defined, e.g. hydrogencarbonate, HCO_3^-, and hydrogen cyanide, HCN, are both classified as inorganic compounds.

oxidation (oxidized species; oxidizing agent) The removal of electrons from a species. The species that has lost the electrons is said to have been oxidized. An oxidizing agent is an electron acceptor as it brings about the oxidation of the species from which it accepts electrons. Oxidation requires both an electron acceptor and an electron donor (species being oxidized); therefore all such reactions are **redox reactions** as both **reduction** (of the electron acceptor) and oxidation (of the electron donor) occur.

oxidation potential The measure of the tendency of a substance to donate electrons (be oxidized) to some electron acceptor. It is the reverse of the reduction potential, i.e. numerically the values are the same but the signs are changed (see **reduction potential**).

phase A homogeneous region in a system, being one of the states of matter. The number of phases is the number of separate physical states in the system, e.g. gas, liquid or solid. Different crystal structures of a substance are also described as being different phases.

photodissociation The breakage of chemical bonds by the absorption of electromagnetic radiation in the visible and ultraviolet portion of the spectrum.

photosynthesis The process by which some organisms (e.g. plants) absorb solar radiation and utilize carbon dioxide and water to produce organic molecules.

plane polarized light   Light that vibrates in only one plane. Light normally vibrates in all planes parallel to the direction of propagation. A molecule containing an asymmetric centre can cause the inclination of the plane of polarization of the light to be rotated.

polar   Having some electrical charge. Used to describe ionic substances or covalent molecules in which the electrons in the covalent bond are unevenly shared (e.g. H_2O; see Chapter 3) so that one part of the molecule has a slight negative charge and another part has a slight positive charge.

polymer   A substance containing a number (often large) of repeating structural segments or sequences. The polymer is formed by the binding together of small units (**monomers**) to form chains. The chains may be cross-linked, giving two- or three-dimensional structures. Natural polymers include proteins and cellulose. Artificial polymers include plastics such as polythene and fibres such as nylon.

precision   A measure of the reproducibility of an observation, i.e. the size of the difference between any one measured value and a number of repeated measurements of the same parameter. The precision is dependent on the size of the **random errors** inherent in a measurement. It is not directly related to the **accuracy** of a measurement as it does not take into account non-random errors.

producers   **Autotrophic** organisms that produce the organic molecules, with their associated energy content, which are then used as a food source by higher members in the **food chain**.

rain out   The removal of gases, dust, etc., from *inside* clouds by incorporation in rain or other precipitation (see **wash out**).

redox reaction   A reaction involving the transfer of electrons from one species to another (see **oxidation, reduction**).

reduction (reducing agent; reduced species)   The addition of electrons to a species. The species that has accepted the electrons is said to be reduced. A reducing agent is an electron donor as it brings about the reduction of the species to which it has donated the electrons. Reduction requires both an electron acceptor (species being reduced) and an electron donor; therefore all such reactions are **redox reactions** as both reduction (of the electron acceptor) and **oxidation** (of the electron donor) occur.

reduction potential   The measure of the tendency of a substance to accept electrons (to be reduced) from an electron donor. The higher the potential in volts, the more readily the substance is reduced. If the reduction potential is positive the reaction is described as being thermodynamically spontaneous, i.e. it is energetically favoured. The value of the reduction potential is dependent upon the concentration of species and the temperature. To aid comparisons, the standard reduction potential is often quoted, having been determined under standard conditions of unit activity (for solutions) or 1 atmosphere pressure (for gases) at 25 °C.

respiration   An energy-generating process utilized by various living organisms in which either inorganic or organic compounds act as electron donors (are oxidized) and inorganic species act as electron acceptors. If dioxygen is the electron acceptor the process is described as *aerobic* respiration. Other electron acceptors include sulphate and nitrate (see Figure 6.4).

sensitivity   A measure of the ability of a technique or instrument to detect small changes. Essentially it is the ability to differentiate between a small real change and a random fluctuation. High sensitivity therefore implies high **precision**.

Sievert   The SI unit that represents the effective amount of radiation energy absorbed by live tissue. The unit takes into account the different types of radiation (α, β, γ, etc.).

$$1 \text{ Sievert (1 Sv)} = 1 \text{ J kg}^{-1} \times \text{radiation quality factor}$$

The previous unit was the rem, where

$$1 \text{ Sievert} = 100 \text{ rem}$$

significant figures   The number of figures in a numerical value, which indicate the **precision** with which a measurement is made or a value is known. For example, a value of 11 has two significant figures, indicating that the value is closer to 11 than to 10 or 12, i.e. it lies between 10.5 and 11.5. Writing the value as 11.1, with three significant figures, indicates that the value lies between 11.05 and 11.15. Often when averaging values or using a calculator to determine a value, the *apparent* number of 'significant' figures is increased. The mean of 11, 12, 11 is 11.3333. However, this should be reported as 11 because the act of calculation cannot increase the precision of a measurement. Similarly the mean of 11.1, 12, 11 would be 11 as the *least* precise figures determine the overall precision. Zeros used to place the decimal point correctly are not counted as significant, e.g. 0.0011 has only two significant figures.

silicates   Minerals based on the $[SiO_4]$ structural unit (see Chapter 8).

solar radiation   Electromagnetic radiation emitted by the Sun. Because of absorption by molecules in the Earth's atmosphere, only a small proportion of the radiation reaches the Earth's surface (see Figures 2.6, 2.8 and 4.14). The energy of the radiation is proportional to the rate of vibration (frequency), with different frequency ranges being given different names, e.g. visible, infrared, ultraviolet (see Figure 2.6).

sorption   The enrichment of one or more components of a particular fluid phase (gas or liquid) in another phase. If the enrichment only occurs on the surface between the phases, the process is called **adsorption**; but if the enrichment occurs in the bulk of the receiving phase, the term **absorption** is used. 'Sorption' is a general term; 'adsorption' and 'absorption' are specific.

standard deviation   A measure of the **precision** of a set of readings. The smaller the standard deviation, the smaller the variation. The formula for calculating the standard deviation can be found in a statistics textbook and it is an automatic function on many calculators. For a **normal distribution**, 68% of all results will lie within one standard deviation of the mean and

99.7% will lie within the range of the mean $\pm 3 \times$ the standard deviation (see Figure 1.7).

stereochemical   Differing spatial arrangement of the atoms or groups in compounds that can affect the chemical and physical properties of the compounds. The compounds are identical apart from these spatial differences about a fixed point.

stereoisomerism   Two or more forms of a compound or other species that can exist because of spatial differences in the arrangement of atoms or groups of atoms either (a) about a plane (*geometrical isomerism*), or (b) about an asymmetric centre (*optical isomerism*).

synergistic   Increasing the effect of two or more components by an amount that is greater than the sum of the individual effects. There is an enhancement of the effects of one chemical due to the presence of other chemicals.

transpiration   The evaporation of water from the leaves of plants (the water having passed through the plant).

wash out   The removal of gases, dust, etc., from *below* clouds in the atmosphere by rain or other precipitation falling from the clouds to the ground (see **rain out**).

wet deposition   Removal of material from the atmosphere by the combined action of **rain out** and **wash out**. Also called 'precipitation scavenging'.

Index

References to figures are indicated by giving the page number in **bold** type. Similarly tables have page numbers indicated in *italics*. Items which appear in the glossary are referred to by the entry G.

266   *Index*